MOUNT ROYAL COLLEGE

D0685888

Date Due		

MOUNT ROYAL UNIVERSITY
LIBRARY

Controversial Bodies

Controversial Bodies

Thoughts on the Public Display of Plastinated Corpses

EDITED BY JOHN D. LANTOS, M.D.

The Johns Hopkins University Press

Baltimore

© 2011 The Johns Hopkins University Press
All rights reserved. Published 2011
Printed in the United States of America on acid-free paper
9 8 7 6 5 4 3 2 1

The Johns Hopkins University Press
2715 North Charles Street
Baltimore, Maryland 21218-4363
www.press.jhu.edu

Library of Congress Cataloging-in-Publication Data
Controversial bodies : thoughts on the public display of plastinated corpses / edited
by John D. Lantos.
    p. ; cm.
Includes bibliographical references and index.
ISBN-13: 978-1-4214-0271-0 (hardcover : alk. paper)
ISBN-10: 1-4214-0271-8 (hardcover : alk. paper)
    1. Tissues—Plastic embedment. 2. Human anatomy. 3. Bioethics. I. Lantos,
John D.
[DNLM: 1. Cadaver. 2. Medicine in Art. 3. Bioethical Issues. 4. Exhibits as Topic.
5. Human Body. 6. Plastic Embedding. WZ 331]
QM556.5.P53C66   2011
611'.018—dc22   2011004503

A catalog record for this book is available from the British Library.

*Special discounts are available for bulk purchases of this book. For
more information, please contact Special Sales at 410-516-6936 or
specialsales@press.jhu.edu.*

The Johns Hopkins University Press uses environmentally friendly book materials,
including recycled text paper that is composed of at least 30 percent post-consumer
waste, whenever possible.

CONTENTS

ACKNOWLEDGMENTS

Many of the papers included in this book were originally presented at a public conference organized by the Center for Practical Bioethics and held at the Kansas City Public Library on December 5, 2008. That conference was supported by the Francis Family Foundation.

GEORGE J. ANNAS, JD, MPH, Chairman and William Fairfield Warren Distinguished Professor, Department of Health Law, Bioethics, and Human Rights, Boston University School of Public Health; Professor of Law, Boston University School of Law; Professor of Socio-medical Sciences, Boston University School of Medicine

CATHERINE BELLING, PHD, Assistant Professor of Medical Humanities and Bioethics, Northwestern University, Chicago

MYRA CHRISTOPHER, President and Chief Executive Officer, Center for Practical Bioethics, Kansas City, MO

FARR A. CURLIN, MD, Associate Professor of Medicine, University of Chicago

JOHN D. LANTOS, MD, Director, Children's Mercy Bioethics Center; Professor of Pediatrics, University of Missouri–Kansas City

CHRISTINE MONTROSS, MD, Department of Psychiatry and Human Behavior, Brown University, Providence, RI

LYNDA PAYNE, PHD, RN, Sirridge Missouri Endowed Professor in Medical Humanities and Bioethics, School of Medicine and Department of History, University of Missouri–Kansas City

GEOFFREY REES, PHD, Instructor, Department of Religion, Health, and Human Values, Rush Medical College, Chicago

TARRIS ROSELL, DMIN, PHD, Rosemary Flanigan Chair, Center for Practical Bioethics, Kansas City, MO; Professor of Pastoral

Theology—Ethics and Ministry Praxis, Central Baptist Theological Seminary, Shawnee, KS; Clinical Associate Professor, Department of History and Philosophy of Medicine, University of Kansas Medical Center, Kansas City, KS

CALLUM F. ROSS, PHD, Organismal Biology and Anatomy, University of Chicago, Chicago

LINDA SCHULTE-SASSE, Chair, Department of Germanic Studies, Macalester College, St. Paul, MN

BARBARA MARIA STAFFORD, PHD, Professor of Art History, University of Chicago, Chicago

NEIL A. WARD, MFA, trade association executive, novelist, and essayist, Washington, DC

Controversial Bodies

Introduction

Plastination in Historical Perspective

JOHN D. LANTOS, MD

M USEUM EXHIBITIONS of chemically transformed, meticulously dissected, and artistically displayed cadavers have become quite popular. Over the last decade, tens of millions of people throughout North America, Europe, and Asia have paid fifteen to twenty-five dollars each to see these exhibitions. The shows are so popular that many museums have had to extend their normal business hours to accommodate the vast crowds of people who yearn to see these unusual specimens. A story on National Public Radio discussed "America's new love affair with corpses on display." These shows have become the most successful exhibits in the history of science museums.[1]

Museums bring science and culture to people. Every day in the United States, 2.3 million people visit museums.[2] There are nearly one billion visits to museums of one sort or another in the United States every year. Many more people visit museums than attend all professional sports events combined. Science museums shape the public perception of new scientific discoveries. So what are we learning from plastinated bodies?

The exhibitions are a clever combination of magnificent technology, naughty sensationalism, and curious artistic aspirations. The technology is the simplest aspect to explain. The displays are made possible by a technique called "plastination." This was invented by a German anatomist named Gunther von Hagens in the 1970s. The technique requires a four-step process that transforms living tissues into moldable plastic. The first step is to embalm the body in a formaldehyde solution to halt decomposition. After any necessary dissections take place, the specimen is placed in a bath of acetone. Under freezing conditions, the acetone draws out all the water and replaces it inside the cells. In the third step, the specimen is placed in a bath of liquid polymer, such as silicone rubber, polyester, or epoxy resin. A vacuum is created, causing the acetone to boil. As the acetone vaporizes and leaves the cells, it draws the liquid polymer in behind it, leaving a cell filled with liquid plastic. The plastic must then be cured with gas, heat, or ultraviolet (UV) light to harden it. A specimen can be anything from a full human body to a small piece of an animal organ.

Plastination was first used to create specimens for the teaching of anatomy in medical schools and veterinary schools. It is still used for this purpose. However, von Hagens also began to use plastination to create stunning anatomic displays that he exhibited to the public. The first show was in Japan in 1995. Von Hagens has become both the world's most famous anatomist and an impresario of anatomy. In addition to his museum shows of plastinated corpses, he has conducted public dissections. In 2002, he rented a theater and performed a public dissection before a sell-out crowd of five hundred people. He has, reportedly, offered to sell plastinated body parts as decorations. Thousands of people have donated their bodies to von Hagens and consented to have their corpses plastinated and displayed. There is now a website for people who wish to donate, explaining how it can be done.[3]

In addition to being popular, the plastination exhibitions have been controversial. In 2003, officials in Munich tried to prohibit the exhibition there, arguing that it violated laws regulating burials and did not respect human dignity.[4] Eventually, the exhibition was allowed to go on but only if some of the more controversial plastinates were removed.

One might have predicted that the exhibitions would have been less controversial in postreligious Europe than in the more God-fearing United States.[5] As Linda Schulte-Sasse explains in her essay, the opposite was the case. While some cities in Europe banned the exhibitions, in the United States the more mild moral controversy seems to have been part of the marketing strategy, rather than a reflection of deep religious or ethical concerns. Many museums that host the exhibitions have created religious and ethics "advisory boards." These boards universally approve of the exhibits or raise minor quibbles about particular pieces within the exhibit (the sensually posed pregnant woman with her uterus dissected and a fetus visible inside is always singled out as particularly troubling and as, perhaps, having crossed some line), which serve only to draw people's attention to the naughtiest or sexiest of the displays. These specimens are then placed in separate rooms, thus reinforcing the implicit message that they are both acceptable and daring.

As a marketing strategy, mild controversy is quite effective. After all, without such controversy, the exhibitions might be as boring as an anatomy class. Or they might have been as marginal and unpopular as the many anatomic museums that already exist, and have existed for centuries, in many American and European cities. Callum Ross and Linda Payne describe some of these museums.

To the extent that the exhibitions are controversial, it is not clear who, exactly, should be protesting them and on what grounds. Should it be, as in Munich, the government? Should it be religious authorities? Should protests come from bioethicists? Anatomists?

The exhibitions have found a moral seam in our society. They are not clearly edifying and not clearly taboo. Instead, they suggestively flirt with cultural taboos in a society that claims not to have any taboos. Part of the popularity of the exhibits derives from the cultural game of "chicken" that they play with moral sentiments and regulatory restrictions. How far can the exhibitors go before someone will step in and try to shut them down? What will be the basis for censure?

A spokeswoman for the Nuffield Foundation, a British foundation devoted to analyzing bioethical issues, suggested one possible basis for censure. "Human tissue should not be bought and sold or otherwise treated as an object of commerce," she wrote. "Body parts, anatomical specimens or preserved bodies should not be displayed in connection with public entertainment or art."[6]

Religious authorities are divided about how to respond to these exhibitions. One rabbi in Germany compared the public display of plastinated corpses to the Nazis' use of human skin to make lampshades. Rabbi Nachum Sauer of the Rabbinical Council of California was not that harsh but wrote that the exhibit does not show proper respect for dead bodies: "Even though it is billed as educational, I feel there are other motives involved than purely scientific, medical ones. I think people may go and see it because of the notoriety—people are always looking for shocking experiences, and it lowers the sensitivities of people in general to the sanctity of the human being. So when people ask me, I recommend that they don't go see it for those reasons." On the other hand, Rabbi Morley Feinstein, a member of the advisory board to the exhibit in Los Angeles, wrote: "I was very proud that the Science Center board and the professionals involved were willing to take the risk of receiving some flak for the higher gain of educating the public about health and the human body. Moreover, when we see with our own eyes the unbelievable design of the human body in all its fine detail, it helps us understand better the Designer who created and shaped humanity."[7]

Catholics are also divided. When an exhibit opened in Vancouver, a representative of the archdiocese objected to the exhibition:

> The concept of the exhibit runs counter to Roman Catholic theology and our belief in the dignity of the human body, which we hold to be created in God's image. Because we hold the body to be sacred, it must be treated with respect at all times. If it is to be used posthumously, the purpose must be worthy of the sacred vessel it is being permitted to use . . . This is a far cry from what we see with the *Body Worlds* exhibit, which includes bodies and parts that were questionably obtained, for use in a popular attraction being widely advertised in the hope of drawing large admission-paying crowds.[8]

On the other hand, representatives of the Archdiocese of Phoenix found the exhibit to have educational value:

> The exhibit's creators have given explicit credit to Christianity with one of the "plastinates" at the beginning of the tour. This figure is "kneeling in prayer"; a nearby placard credits our 16th Century popes with first encouraging the study of anatomy to increase medical knowledge of the body. This is appropriate, then, to see the viewing of the exhibit as being an outgrowth of the Church's perpetual concern to increase true knowledge in God's creation.[9]

Opposition from religious groups is, of course, good for business. As Mark Twain noted long ago, the best thing that could happen to one of his books was for it to be banned somewhere. Sales would always rise immediately and dramatically.

Another moral concern surrounding the plastination exhibits focuses upon the source of the bodies themselves and how they were obtained. A report on the television news show *60 Minutes* suggested that the bodies were obtained without consent and that some may have even been the bodies of executed political prisoners in China. These allegations first surfaced in a 2004 article in the German newsmagazine *Der Speigel*, which reported that some of von Hagens's corpses had bullet holes in the backs of their necks.[10] Since then, an investigation by reporters on the ABC television news show *20/20* has suggested that many corpses were obtained without consent.[11]

In New York, concerns about the origins of the corpses led State Attorney General Andrew Cuomo to investigate the company, Premier Exhibitions, that was sponsoring the exhibition there. (Premier Exhibitions is not associated with von Hagens.) Cuomo concluded that consent for the use of the bodies had not been obtained and ordered the posting of a sign stating that "this exhibit displays human remains of Chinese citizens or residents which were originally received by the Chinese Bureau of Police. The Chinese Bureau of Police may receive bodies from Chinese prisons. Premier cannot independently verify that the human remains you are viewing are not those of persons who were incarcerated."[12] Customers who could establish that they would not have attended the New York City exhibition had they known of the questions about the origins of the bodies were eligible for refunds.

Anatomists, too, were conflicted about how to respond to the exhibits. The German Anatomical Society censured *Body Worlds* as severely violating the educational, scientific, and ethical principles of its society.[13] The Anatomical Society of Great Britain and Ireland and the British Association of Clinical Anatomists expressed their concern that *Body Worlds* would sensationalize and trivialize human dissections and might lead to a decrease in the donation of bodies to medical schools or for organ transplantation.[14] As with religious groups, the professional anatomists are not unaimous in their condemnation. New Zealand anatomist D. G. Jones noted, "Even those who wish to criticize the *Body Worlds* side of plastination would do well to remember that von Hagens has also opened up exciting new vistas for research."[15]

The Long History of Controversies about the Study and Teaching of Anatomy

Debates about these exhibitions are part of a long debate about the study and teaching of anatomy. The debate focuses on whether there are any appropriate prohibitions on the uses to which dead bodies or body parts can be put for medical therapy, medical research, and medical education. That debate raises questions about the proper limits of scientific inquiry into bodily things.

There are, of course, good and important reasons to study anatomy, physiology, and pathology. Such study is the only way to understand how disease affects human beings and how treatment might help or harm us. We must use the bodies of both animals and people as experimental or educational objects in order to learn and to teach doctors to be doctors. Anatomic study, by itself, is neither right nor wrong, but it always raises issues about the ways in which it can dehumanize us. Farr Curlin discusses these issues.

The modern history of anatomic study is inextricably tied to the modern history of medicine and, in particular, the development of formal schools of medicine. The earliest such schools in Europe were in France and Italy and were founded in the thirteenth and fourteenth centuries. Katherine Park recently reviewed the origins of dissection in Europe. She effectively destroys the myth that the Catholic church opposed dissection in the pre-Renaissance era. She writes, "From at least the early twelfth century, opening the body was a common funerary practice. Over the course of the fourteenth century, it also established itself in Italian medicine as not only tolerated but frequently requested on the part of individuals and their families. Not until the mid-sixteenth century do we begin to see persistent hints of a new popular suspicion concerning dissection."[16]

Using primary sources from Italy and France in the fourteenth and fifteenth centuries, Park shows how modern concerns about anatomy, anatomists, and grave robbers arose not because of the objections of religious leaders but because of the rise of science, the new prominence of anatomic study, and the associated rise of grave robbing. This, in turn, led to new burial rituals, designed in part to protect the dead bodies of loved ones and foil the grave robbers. According to Park, "This suspicion [of dissection] was not rooted in age-old taboos; rather, it grew out of dramatic new anatomical practices widely perceived as violating not the sanctity of the body, in the first instance, but the personal and familial honor expressed in contemporary funerary ritual. And it was reinforced by new, and not unwarranted, fears that anatomists themselves occasionally acted as executioners." In short, she argues, it was the hubris of science and medicine in seeking more and more bodies to study, rather than religion's taboos, that led to public opposition to human dissection.

The most famous anatomist of the period was Andreas Vesalius, whose great work, *On the Fabric of the Human Body* (1543), revolutionized our understanding of body structure and function. Vesalius was born in Brussels, studied in Paris, and then joined the faculty of the medical school in Padua. His anatomic knowledge came from the numerous dissections that he carried out himself. The corpses that he dissected were obtained by robbing graves. He himself, and the students who worked with him, would regularly raid not just graves but even the homes of the newly dead where the fresh corpses were laid out prior to burial, steal them, and swiftly dissect them.

Dissections in sixteenth-century Europe were often public events that attracted large and often raucous crowds. Anatomic study of human cadavers was legalized in Amsterdam in 1555, although it was limited to male bodies. (The public dissection of female bodies would not be accepted until one hundred years later, in the late 1600s.) The pressure for legalization of human dissection came from the surgeons' guilds. The Dutch government resisted the blanket legalization of dissection and opted, instead, to authorize a limited number of public dissections. They appointed one doctor to the post of Praelector Chirurgiae and Anatomie, or City Anatomist. This man had the right—and obligation—to dissect a certain number of corpses per year. The law stated that at least one of these dissections should be open to the public. Annual public dissections were conducted in large theaters. The first few rows were reserved for magistrates and members of the surgeons' guild. The general population was welcome to purchase tickets for the remaining seats. A poem written in 1646, by Caspar Barleaus of Amsterdam, suggests the mood of the observers—or at least the mood to which one might aspire:

Here lies spread out Man
And offers to all the World spectacles of his pitiful state.
Brow, finger, kidney, tongue, heart, lung, brain, bones, hand,
Afford a lesson to You, the living.
You may examine here publicly that which is healthy
And that which is diseased.
While you contemplate the remains of the defunct,
Learn that it is through God that you live hale,
Teach this to yourself.

Public dissections became the subjects of paintings by leading Dutch painters, including Rembrandt. The style of the paintings, as historian Julie Hansen points out, is that which Dutch painters used to commemorate large-scale historic events. The style—and the very fact of the paintings—suggests a civic pride regarding public dissection. The Dutch, perhaps as they do today, were willing to acknowledge and legalize what was illegal elsewhere. Today, in Amsterdam, hashish, prostitution, and euthanasia are all legal. In those days, it was public dissection.

There are detailed descriptions extant of the public dissections. They were usually the culmination of a month of public education—lectures by prominent physicians about medical topics. Microscopes were demonstrated, experiments were performed on live animals, and finally there was a public dissection. The anatomic theater was lit by scented candles. A flautist or harpist played background music. After the dissection, there was a banquet for the surgeons' guild, paid for by the profits from entrance tickets.

The bodies for public dissections were usually those of executed criminals or sometimes of transients who had no local relatives to claim their bodies. The law stated that the subjects for public dissection could not be citizens of the city where the lesson would take place. It was also stipulated that the surgeons' guild had to pay for a proper church burial for the subject following the dissection.

Anatomy theaters also housed anatomy museums. The largest of these was at Leiden, where the anatomists created a series of remarkable artistic displays using body parts and images to connote the evanescence of life—hourglasses, extinguished candles, wilting flowers, and the like. Public dissectors often saved the best specimens for display in their anatomic museums.

Many of the exhibits in the Leiden Anatomical Museum were created by one remarkable man, Frederic Ruysch, whose work comes as close as that of any historical figure to the work of our modern-day plastinators. Ruysch lived from 1638 to 1731, studied medicine at Leiden, and developed a new technique to preserve and display anatomic specimens. His technique was a closely guarded secret but, according to recent historians, probably involved the use of talc, wax, oil of lavender, and cinnabar.[17] He used colored pigments to highlight different anatomic structures and preserved his specimens in a preservation solution of brandy and black pepper.

Ruth Richardson, in her book, *Death, Dissection and the Destitute*, traces the history of the commodification of the corpse in Britain. She shows that grave robbing was common even in Shakespeare's day, as shown by his epitaph (written in 1616):

Good friend, for Iesus sake forbare
To digge the dust encloased heare.
Bleste be ye man [that] spares these stones,
And curste be he [that] moves my bones.[18]

By the late seventeenth century, dissection was so common that the dead human body began to be bought and sold like any other commodity. Richardson cites a 1728 publication noting that "the Corporation of Corpse Stealers, I am told, support themselves and Families very comfortably."[19]

In 1752, British judges were first permitted by law to condemn murderers to dissection, rather than gibbeting in chains. Dissection, at that time, was seen as further punishment, and family members of a condemned murderer often tried to rescue the body of their loved one from the gallows. Attempted rescue of the corpse from the surgeon was punishable by banishment to the American colonies for seven years.

In 1768, William Hunter was appointed professor of anatomy at the Royal Academy of Arts. He used this post to attract students to his own private anatomy school in London, where each student had access to an individual corpse. Hunter was also one of the first surgeons to base his surgical technique on detailed study and knowledge of anatomy and to teach students to do the same. He also started his own anatomic museum. Where did Hunter get his corpses? He needed far more than would have been obtainable from the gallows and so had to turn, like anatomists before him, to grave robbers. In some cases, the grave robbers were the medical students themselves. In Scotland, anatomy students could pay for their tuition with corpses rather than cash.

The corpse had a peculiar legal status. It was not property and so, technically speaking, could not be stolen. When a grave was robbed, the crime was, technically, against the grave, not against the corpse. Grave robbers, then, were not thieves. They were desecrators of property. There was a divergence, though, between the law and the public's perception of the nature of the crime. Families clearly saw grave robbers as equivalent to thieves.

According to Richardson, by the early nineteenth century there was a fairly stable "market" in corpses, with the going rate in London being around eight guineas. At that time,

> Corpses were bought and sold, they were touted, priced, haggled over, negotiated for, discussed in terms of supply and demand, delivered, imported, exported, transported. Human bodies were compressed into boxes, packed in sawdust, packed in hay, trussed up in sacks, roped up like hams, sewn in canvas, packed in cases, casks, barrels, crates, and hampers, salted, pickled, and injected with preservative. Human bodies were dismembered and sold in pieces, or measured and sold by the inch.[20]

Similar grave-robbing practices and markets developed in the United States and led to a fairly well-developed partnership between grave robbers and medical schools in Boston and Philadelphia. Michael Sappol's marvelous book, *A Traffic of Dead Bodies: Anatomy and Embodied Social Identity in Nineteenth-Century America*, traces the development of elaborate networks of supply and demand in these American cities. He explains that, "as black markets in cadavers flourished, so did a cultural obsession with anatomy, an obsession that gave rise to clashes over the legal, social, and moral status of the dead. Ministers praised or denounced anatomy from the pulpit; rioters sacked medical schools; and legislatures passed or repealed laws permitting medical schools to take the bodies of the destitute."[21]

The commercialization of the body was not limited to medical education. Museum exhibits—precursors of today's plastination exhibits—thrived in nineteenth-century London. Such museums generated many of the same controversies as do museums today.

In a paper about public anatomy museums in nineteenth-century Britain, A. W. Bates describes how such museums were sometimes shut down by the authorities under the terms of the Obscene Publications Act.[22] He begins with the story of one particularly popular exhibit, the Kahn Anatomical Museum. On December 18, 1873, the magistrate ordered that the wax anatomic models that had been on display in the museum be destroyed with a hammer.[23]

Bates argues that anatomy museums became popular through their advertised promise of being both educational and pious. Medical anatomy teachers and museum proprietors argued that, by promoting self-knowledge and revealing created order, anatomy was in fact an argument against atheism. In such museums, the general public could learn about anatomy without going through the unpleasantness of actually dissecting a cadaver. By the 1850s, he shows, the museums had changed their mission. Museum proprietors began dispensing medical advice and offering treatments for venereal disease. Bates writes, "Horrifying models of diseases alarmed patients and entertained casual visitors." Eventually, the law came to treat such exhibitions not as education but as obscenity. Anatomic museums were closed, and anatomic displays were taken out of the public domain and restricted to the domain of medical education. "Professionals," Bates notes, "by virtue of their education, social background and character, were deemed impervious to influences that could corrupt the weaker-minded public."

Modern Echoes of Ancient Concerns about Anatomic Displays

Von Hagens and his modern imitators are clearly working in the tradition of Ruysch, Hunter, and other purveyors of public anatomic displays. Von Hagens, like Ruysch, claims that the exhibitions are educational. As he writes, "My work continues the scientific tradition whose recurring theme is that research should serve the general enlightenment."[24] He regards the dissection arenas that opened in Padua in 1594 and Leiden in 1597 as predecessors of his exhibitions. It is, perhaps, not surprising that exhibitors of plastinated bodies face problems similar to those faced by other anatomists who crossed the line between scientific study of anatomy and popularization.

While public displays of anatomy are not new, these particular displays may have new meanings because they take place at a time when biotechnology is transforming many of our ideas about the human body. Debates about anatomic displays must be placed in the context of debates about other uses of human bodies and tissues. These include questions about stem cells, reproductive cloning, organ transplantation and organ sales, in vitro fertilization, and the patenting of genes and even species. These developments in biology change the way we think about our fundamental nature as embodied creatures. Plastination is not a unique or even a uniquely disturbing example of these developments, but it is a uniquely popular one. Thus, it offers a lens through which to examine a complex set of questions raised by advances in synthetic biology, medical imaging, and even religious thought about the body.

The popularity of plastinated body exhibits masks a complex mix of emotions. We admire the artistry of the exhibitions. We visit the exhibitions because we yearn to learn more about ourselves, about how our bodies are put together and how they do the marvelous things that they do. We also go to the exhibitions for titillation. Anatomic museums have always flirted with and skirted the lines between education, entertainment, and prurience. They are one of the few places where it is permissible to gaze at naked bodies.

The admiration is both for the magnificence of nature and for the clever blending of religion, art, and science. The artistic elements of the exhibitions are mostly in the cleverness of the display of individual specimens. The plastinators seem to be sculptors who have invented a new medium using the raw materials of the dead human body and the chemical processes that they have invented and patented. These techniques essentially turn the corpse into a sort of putty that can be sculpted and molded into shape and then hardened and fixed in that shape. The result is a seamless hybrid of the organic and the inorganic, of that which was made by nature (or God) and that which was made by humans. There is wonder to be found in the work of both creators, the stunning intricacy of the various body systems—muscles, nerves, bones, blood vessels, one overlaying and intertwined with the other, all synergistic and dependent upon one another. Analogies to complex machinery are unavoidable and always inadequate. The machinery of life is, inevitably, mechanistic, but it seems to be something more. It is also artistic.

Religious awe feels both inappropriate at such a gaudy commercial venture and also inevitable as one gapes in wonder at what seems too intricate to be the work of chance alone. Could all of this really have been made without a maker? How many false starts, how many years, how many events, each so improbable as to be, for all practical purposes, impossible, would have had to have taken place in the primordial soup to get just this delicately balanced organism? Or these organisms, really, since the plastinated body exhibitions include not just humans but also horses. (In fact, plastination is used in schools of veterinary medicine to teach animal anatomy. Many schools have developed "museums" of plastinated animal models as study aids for their students.)

Along with the feelings of admiration and awe, almost as part of them, come feelings of disgust or unease. The exhibitions seem tawdry, exploitative, sensational, even pornographic. Part of their fascination is the bizarre nudity, the fact that the bodies are, well, just bodies. Along with hundreds of other museumgoers, we stare at the women's breasts and the men's penises with a mix of clinical detachment and titillation.

The study of anatomy has always been both essential and ghoulish. Moral and religious debates about the appropriate and inappropriate uses of the body almost always end in tentative compromises—we do not want to turn the body into a commodity, but we do feel an obligation to learn from the body.

In this book, we try to understand the ways in which the current craze in plastination and the commercial success of the museum shows interact with ideas about medicine and medical education and the appropriate viewing of the dead body and its liminality in culture. We do so through essays that bring different voices and different perspectives—some more scholarly, some more personal—to analyze the basis for the moral and aesthetic judgments that we make about these exhibits. We try to address some of the similarities and some of the differences between exhibitions of plastinated bodies and anatomic exhibitions in the past. We speculate about the goals and the limits of public education about anatomy.

Catherine Belling, a professor of medical humanities at Northwestern University, and George Annas, a professor of health law at Boston University, examine the hopes and dreams of people who might decide to donate their body to a plastination exhibit. Geoffrey Rees, a theologian and bioethicist, examines the ways in which plastination exhibitions seem more like a throwback than like an advance. He speculates about what we might see in anatomy museums in the future.

Two physicians, Christine Montross, a psychiatrist at Brown University, and Farr Curlin, an internist at the University of Chicago, reflect on the similarities and differences between a medical school anatomy class and an anatomic display in a museum. Callum Ross, who teaches anatomy to first-year medical students at the University of Chicago, discusses the ways in which the educational value of public anatomic displays could be improved. Lynda Payne, a nurse and historian of medicine, wonders what William Hunter, a pioneer of anatomic research, might have thought of plastination.

Linda Schulte-Sasse, who teaches German and German Civilization at Macalaster College, writes about the differences in how von Hagens was received in Germany compared to the United States. Tarris Rosell, an ethics consultant at Kansas University Medical Center and professor of moral philosophy at the Baptist Theological Seminary in Kansas City, discusses some Christian concerns about the public display of plastinated cadavers. Myra Christopher, director of the Center for Practical Bioethics in Kansas City, tries to understand her own moral qualms about going to see the plastinated body exhibitions.

Finally, two essayists examine these exhibits as art rather than as science or public education. Barbara Maria Stafford, an art historian whose research and writing focuses on the many ways that technology has changed artists' perceptions of bodies and of what it means to be embodied, relates plastination to broader aesthetic and cultural movements of our time. Neil Ward, a novelist and association executive based in Washington, DC, wonders how plastination fits into the tradition of twentieth-century conceptual art.

We hope that the essays in this book might help the inquisitive or concerned museumgoer who exits a plastination exhibition and wonders, "What did I just see, and what does it mean?"

CHAPTER ONE

Being Non-biodegradable

The Lonely Fate of Metameat

CATHERINE BELLING, PHD

There's a great future in plastics. Think about it.
MR. MCGUIRE IN *The Graduate* (1967)

THE PRESERVATION OF HUMAN BODIES by plastination converts humans into objects. These dead people do not appear to decay, and so we are protected, despite the graphic exposure of their insides, from the physical revulsion we might expect to feel in the presence of human remains. Whether we understand these objects as biological specimens or as aesthetic creations, or both, their explicit resistance to decomposition tells us, on a visceral level, that they are something other than human remains. As such, they are also no longer subject to the special regard that is always entailed by horror.

Many objections to the *Body Worlds* exhibitions seem to turn on the distinction between science and art.[1] If we think of the plastinates as specimens, we imagine observing them dispassionately and with control. If we imagine them as works of art, we risk acknowledging the pleasure that looking at them makes us feel, and such pleasure might remind us of the horror it has replaced. Instead, we try to focus on the improving effects of our seeing. It is morally justifiable, the argument in its crudest form goes, to turn a human being into a dissected display in order to further objective factual knowledge about the body and to disseminate that knowledge to the public. Democratic science education is a legitimate, indeed required, by-product of science. A more pragmatic justification rests on the public health benefit that Von Hagens claims will follow from seeing the causes of death in the bodies themselves. This revelation is expected to promote a self-policing inward gaze in viewers who, having seen the inner lesion—the smoke-blackened lung or sclerotic artery—will, it is argued, work harder at their own wellness disciplines. Donors are expected to become posthumous object lessons.

But plastinates are never, strictly speaking, posthumous. They have foregone the transition of burial—the "inhumation" from which "posthumous" takes its literal meaning—or equivalent sociocultural rituals marking the end of a body's life and situating that body, as a material object, apart from the physical space of the living. Loved ones will go to great lengths to retrieve the material remains of their dead in order to dispose of those remains appropriately, to enact the performative acknowledgment both of what they have become and of what they were before. A memorial service without a corpse is not the same thing as a funeral. It is about memory and closure, but not about matter. Knowing their bodies will be transformed into museum exhibits, how do the loved ones of plastination donors expect to remember them?

Plastination elides the cultural signs we attach to human remains. We exclude corpses from the presence of the living, abjecting them— casting them out—and simultaneously marking them as special, as sacred. Plastinates instead are allowed to remain, uncovered, in the physical presence of the living. We move among them, and no matter how sincere our admiration or reverence or absorption in what we can learn from them, we move on. We are not petrified at the sight of them. They take on, then, something of the character of other inanimate objects that occupy and decorate the spaces of the living. If we are not offended by their display, it is because we have taught ourselves to think of them as things.

The usual process of marking a person's becoming posthumous attends closely to the fact that the dead change, becoming increasingly repellent. In the West, we accommodate that process either by burying the corpse or by burning it (arguably a highly accelerated process of decomposition). Both interment and cremation are predicated on managing and then disposing of a body that will rot. Plastination might appear to be a simple advance on those funerary practices, but the endpoint is different.

Born and raised outside the United States, I am perplexed by the custom of embalming and cosmetizing the dead and displaying them in elaborate coffins before they are buried or cremated. The practice appears to deny the reality of death but, worse, it seems grotesque both in its effort to conceal the physical effects of being dead and in its inevitable failure fully to achieve this cover-up. Decay is a part of death (except for the supernaturally holy or evil), the dynamic process of biodegradation that marks the body as having-been-alive. The painted, embalmed corpse never looks like the sleeping live person. This last gesture at the end of life seems inevitably both false and hopeless. Or so it seems to me. But perhaps this is less a cultural difference than an idiosyncratic one. Ever since I can remember, I've imagined being buried in the garden in no more than a shroud and allowed to turn as quickly as possible into worm food and soil. No doubt this is because I remember more than one family pet being buried, unceremoniously but never insensibly, in a good deep hole under our syringa tree. After embalming, the buried corpse still decays, but slowly and toxically, seeping chemicals and organic matter onto the expensive satin lining of a heavy, sealed coffin. This seems like a tortuously unnecessary kind of purgatory, but despite the resistance or denial that motivates it, the practice must submit finally to the fact that the dead always eventually become unfit for human company. This is not the case with plastination.

One might say that plastination is at least more honest than embalming is about the desire to keep human bodies from becoming food for lesser creatures. And the stasis that plastination aspires to is not the illusion of living sleep but a new form, a frank exposure of what decomposition rather less meticulously uncovers. The plastinated corpse's new form is at once familiar and alien to those who view it. If forced to piece together the mental inner anatomies we accumulate over a lifetime of biology classes and TV medical documentaries and visits to the doctor, we might imagine something not unlike a plastinate—except that, unlike the self we imagine exposed by surgery or trauma or scavengers, this new form is dry. It does not ooze or seep. Its organs are not shiny with defensive mucus or dripping slick blood. Nothing overflows. Instead of bloody red and coagulated black, the bodies are tasteful, reassuring cream and terracotta. Plastinated bodies look rubbery and reductive, with an odd, furry fiberishness, as if woven from very old hair. They are not fresh meat.

Yet in fact they are—partly—meat. They are about 30 percent biomass, about 70 percent plastic, silicon rubber, or epoxy resin or some other synthetic product.[2] DNA-bearing protein is amalgamated with plastic at the level of the cell into a hybrid in/organic material that has the genetic properties of flesh and the inedible staying power of polyurethane.[3] Those who donate their bodies to Gunther von Hagens apparently aspire to this condition, to being reconstituted out of reach of decay even while retaining the solid matter of themselves. Not just models for Von Hagens's sculpture or armatures to support it, they are the statues themselves. Their biotic matter is wholly captured by and assimilated to the plastic. Yet while organic matter saturates the synthetic, what makes the plastinate "lifelike" is not its composition but an optical illusion based on its presentation as an object, in the tension of a pose or the arrested trajectory of a familiar gesture. Without the information that they began as human beings, we would find that in plastinates the human—though never the individual human donor—lies not in the solid flesh but in the semiotics of representation. No matter how anatomically informative they may be, they are artifacts fabricated from human raw materials. Who, then, would choose to become such a thing?

Robert Ashton is a journalist who writes for the British *Independent*. He is also a body donor. After becoming a donor, he was invited to visit the plastination factory in China. He really does call it a factory, this place he plans to send his own corpse. One wonders at the symbolic global-economic implications of the West shipping its dead to China as raw materials and receiving them back from its factories as plastic. Ashton observes a fresh intake of cadavers and finds that, before processing, they are disconcertingly recognizable as real people, with visible signs of the identity that flaying and dissection will soon erase: "These muslin-wrapped dough-skinned corpses are somewhat less attractive than the finished product. Quite simply, they look dead—which, of course, they are. Worse, perhaps, is that they look like the people they until recently were."[4] For Ashton, these bodies are unattractive less because they are dead than because they are human. In order to be attractive—aesthetically pleasing—they must be stripped of their skin and with it the signs of individuality that mark them as persons rather than manufactured products.

Yet Ashton does not seem to mind that, in being made attractive, he will also have to be stripped of identity. His motivation is positively exhibitionistic: "No rotting in the ground for me, or quick release as a puff of smoke from some crematorium chimney. I wanted to be here, with the other exhibits; skinned, posed and proud."[5] Filling out the paperwork, he notes that he must give specific permission to be part of a public show and welcomes "the opportunity to spend perhaps 100 years showing the bits to all and sundry that I'd never quite plucked up the courage to display on Holkham Beach, a well-known nudist haunt an hour's drive from my Norfolk home." Ashton does not seem to be thinking of his future incarnation as an anatomy lesson or a cautionary display of bodily errors. He expects to be admired for his appearance.

I don't find the plastinates beautiful, exactly. That odd hairiness is off-putting. They are impressive, certainly, in the way a great deal of unbeautiful conceptual art is impressive. And also thought-provoking. Ashton, though, is not exactly thinking of himself as gallery art. His aesthetic values are distinctly popularist: plastinates are beautiful because they are thin. This is a flattering effect of the chemicals' removal of fat as well as water—or, as Ashton puts it, "Spare tyres, bat wings and other forms of blubber stacked through decades of self-indulgence simply dissolve in vats of acetone." This is, of course, a rhetorical performance, a daring challenge to those whose sensibilities are easily offended. At the same time, Ashton is speaking the contemporary conviction that living human bodies are evaluated according to their fitness for display. Not in shape for Holkham nudist beach, Ashton imagines securing himself a very long future as a good-looking—and globally looked-at—corpse. His motivation may not be unique.

But as we know, the long-term display, even the self-display directed in advance, of human remains is at odds with the careful boundaries humans universally erect between the living and their dead. Only very special exceptions—saints, Lenin, traitors in gibbets—have historically been allowed to transcend these. The argument that plastinates do not decompose, are not nauseating to most people, does not seem enough to allow them visibly to inhabit the rooms of the living.

We might compare them with a very different kind of unburied corpse. Jim Crace's novel *Being Dead* recounts the lives of two people who end up murdered on a remote beach. Parallel to their biographies, Crace narrates the dynamic biological afterlife of their bodies, tracking in poetic detail the effects of gravity, insects, weather. For instance, on "their fourth day of putrefaction":

> They had both dissolved and stiffened. They were becoming partly semi-fluid mass and partly salted drift; sea things. They even smelt marine, as corrupt and spermy as rotting bladder-wrack or fish manure.[6]

This is an inhuman scenario, and potentially horrific. And there is no inhumation—at this point in their decomposition, at least, nobody knows these two are dead. Nobody knows. This fact leaves them in a state of indistinct continuity with the natural materials and creatures around them, and they are for this very reason not identifiable as human, except from the imaginary vantage point offered by Crace's omniscient narrator. This all changes, though, when they are discovered and the beach becomes a crime scene. A tent erected to keep off the sun, the maggots "hoovered up," they enter a temporary—and repellent—stasis before the usual funerary practices are brought to bear and they are safely resituated across the appropriate boundaries. For Crace, his telling of their past lives and his telling of the beginning of their corporeal afterlife both belong in the liminal (and littoral) zone they occupy before becoming posthumous.

Plastination, we imagine, produces bodies that will never stink and rot. Natalia Lizama, describing the melancholy "post-biological" identity of the plastinate, sees the plastinate in the way Von Hagens encourages us to, as "endowed with a form of synthetic immortality; it will not decompose or disintegrate but will retain its vivid, animated appearance and pose. Through plastination the body is corporeally resurrected and suspended indefinitely in a state of posthumous being somewhere between organic death and synthetic life."[7] Yet the arrest of decomposition is not resurrection—it reverses nothing—but instead holds the material body in an artificial freeze-frame outside the dynamic processes of biology that extend always beyond the limits of the lifespan. And Lizama is wrong, it appears, about its immortality. Its material may no longer be biodegradable, but even plastinates are perishable.

The following description in a blogger's account of visiting a *Body Worlds* exhibit reminds us powerfully of this. She observes that the specimens "not encased in glass seemed dusty, and the plastinated artery systems had broken bits in their cases."[8] The signs of efforts to postpone this degradation make the bodies resemble furniture, tchotchkes: "The funniest things, to me anyway, were the taped cross-sections of the human body. Since they are so fragile, I'm sure cracking and breaking is common. But I could *see* the tape, and it pretty much just looked like packing tape." The materiality of the objects as things reveals, to the attentive eye of one disposed to be amused rather than offended, the complete demystification of their humanness, a demystification that belies Von Hagens's rhetoric of awe and of permanence.

The process of decomposition is not particularly slow, either. "I realize," she says, "that there is now a *Body Worlds* 2 and 3, so most of the specimens in this collection were older (made from 1995 and up) and you could tell some of them had seen better days." The writer does not, we assume, mean the days they saw when they were alive. Those days are already excluded from the trajectories of these objects. They date, now, only to the days when they were newly manufactured, not yet dusty or cracked. These body-things seem, if one allows oneself to reanimate them for a moment, almost intolerably lonely. What will happen if nobody wants to pay to visit them anymore?

Robert Ashton is aware of the rather perverse sociality of his decision to be memorialized in an exhibition catalog rather than on a gravestone and to inhabit museum halls rather than the family plot:

> The difficult part was going to be explaining to my wife . . . that I might not be keeping a double grave warm for her at Blythburgh, that charming church overlooking the Suffolk coastal marshes. No, she'd have to rest there alone as I was going off to be paraded around the world. She would have to rot alone.[9]

In the end, it seems, the parade will be less splendid than he imagines. In becoming a manufactured product he will become only as permanent as all our products, even the most resilient of our polymers. Rather than human remains, he has chosen eventually to become garbage.

CHAPTER TWO

Lifelike Humans

Playing Poker with James Bond and Ted Williams

GEORGE J. ANNAS, JD, MPH

> Surround yourself with human beings, my dear James. They
> are easier to fight for than principles . . . But don't let me
> down and become human yourself. We would lose such a
> wonderful machine.
>
> MATHIS TO BOND IN IAN FLEMING'S *Casino Royale*, 1953

I N THE MOST RECENT MOVIE VERSION of *Casino Royale* (2006), Daniel Craig, who plays British agent James Bond, follows an international criminal into a *Body Worlds* exhibit. The exhibit, featuring three plastinated bodies playing poker, is the perfect backdrop to foreshadow the high-stakes poker game around which the movie's plot pivots. The exhibit of plastinated human remains also comments slyly on the ethical environment of the "oo" British agent with a "license to kill." Bond transforms living people into corpses; Gunther von Hagens transforms corpses into inorganic sculptures. Bond's knife is a death-dealing weapon; von Hagens's plastination process, performed after skinning the corpse with a knife, turns corpses into facsimiles of frozen life. The Daniel Craig Bond is in possession of a newly minted "license to kill." The two-minute sequence invites the ethically complex question: by what authority—ethical, civic, or otherwise—did Dr. Gunther von Hagens obtain his license to plastinate human remains?

The Bond movie highlights another feature of pop culture: torture has become a common theme. Bond himself is viciously tortured near the end of *Casino Royale* and is rescued just before his penis is to be amputated. Torture provided the plot lines for one of the most popular post-9/11 TV series, 24, with its Bond-like hero, agent Jack Bauer. In one of the world's best-selling books, *The Girl with the Dragon Tattoo*, Lisbeth Salander is tortured by villainous representatives of two of the world's major professions: law and medicine. At least two of von Hagens's figures, the Runner (1997) and the Skin Man (1997), appear as if they could have been victims of a torturer who literally skinned them alive. In all of these representations, art seems to be trying to tame torture and transform it into part of the unremarkable background of the culture.

Perhaps, after twenty-one Bond films over a half century, we are desensitized to Bond's license to kill and accept it as a logical and necessary adjunct to a secure world order. Are we prepared to take as equanimous a view of von Hagens's license to plastinate and display corpses? That two-minute cameo appearance of the *Body Worlds* exhibit in *Casino Royale* suggested that perhaps we have arrived at just that point (or else it is simply one of the best "product placements" in film history).

A 2006 press release from von Hagens about the use of the exhibit in the film (set in Miami in the movie but actually filmed in Prague) did not focus on the bodies at all. Instead, it talked about the ways in which Gunther von Hagens, as a youth in East Germany, was inspired by the anti-communist messages in James Bond movies. In his words, Bond "stood for the power of the individual against communism and was anti-authoritarian and unconventional . . . He was very hardworking. He was always on duty, lived only for his mission, and used all his abilities to realize his mission."[1] The only Bond film von Hagens references in the press release is *Goldfinger*, possibly because the film is associated with an image of the woman who is killed by being spray painted with gold over her entire body— objectified in a way suggestive of plastination.

Complementing this apparent quest to desensitize our experience with pain and mortality, von Hagens addresses another deep-seated need: our need for, and fascination with, a sense of authenticity. Complex rituals to preserve corpses are not new. They are at least as old as the elaborate embalming methods used by the ancient Egyptians. Nor is the creation of images of corpses posed as live people new, as, for example, Vesalius's *Muscle Man* (1543) illustrates. Von Hagens justifiably designates Vesalius the "founder of the science of anatomy" and takes much of his own inspiration from Vesalius— who was also the first person to construct a human skeleton made of "real bones."[2] As science and technology have vastly improved our ability to explore our bodies "under the skin," as von Hagens puts it, we have wanted more. We want to learn what goes on in there. Anatomic studies tell us about ourselves. Pathologists have taken the dissection pioneered by anatomists to a much higher (and less destructive) level, and use it for a different purpose: to determine the cause of death, thereby letting the living learn from the dead.

Both the title and the subtitle of Gunther von Hagens's exhibitions of plastinates state the core of the discussion he has provoked: *Body Worlds: The Anatomical Exhibition of Real Human Bodies.* The word *bodies* is repeated, the second time with its important attributes: "real" and "human." The phrase "anatomical exhibition" seems distinctly of secondary importance, although it is clear that von Hagens wants to be seen in the context of the great anatomists of the past. In his emphasis, von Hagens is insightful: people do not come to learn anatomy. The presence of *real* human bodies—neither models of human bodies nor animal bodies—is what evokes the intense emotions that draw people in.

The interplay of life and death, corpse and "real body," is never far from the surface in *Body Worlds.* It draws us to the exhibit and provides the same attraction-repulsion emotions we have toward our mortality and our own deaths. Plastination puts the corpse on display in a way that seems to provide us with a new alternative to cremation and burial (which hide the corpse from view), giving us at least the illusion of greater control over our bodies after death. *Body Worlds* raises questions about what it means to be real or authentic. In its application of technique to the human body, *Body Worlds* almost seems too good to be true, too well constructed to be real.

How do we define the authentic and real? The caves at Lascaux, in France, contain paintings that are more than seventeen thousand years old; the actual caves are off-limits to tourists (to protect the paintings from bacterial influences); one can, however, visit a duplicate of two of the caves containing the major paintings. Assuming that the duplicate caves really are indistinguishable from the real caves, is the experience for the visitor different in quality? If not, why would it matter whether the plastinated bodies are made from real humans or are synthetic facsimilies, in terms of their scientific or artistic value? What is the nature of the appeal that "authenticity," in itself, conveys?

Of course, we know that movies are fantasies (there is no 007 or Jack Bauer in real life), and movie sets are constructs designed to make us suspend our disbelief and pretend that the resulting images are "real," at least until the movie ends. But how does the quest for authenticity apply to the plastinates, and what does it mean to be an authentic corpse? Does the adjective *lifelike* state a paradox or an affirmation?

Von Hagens concedes that viewers can see his posed plastinates as art, and often do. In this regard, *Body Worlds* can be viewed as a natural extension of the pop art movement in the United States, which used found objects in its execution and quickly adopted the images of American advertising (e.g., Andy Warhol's Campbell's tomato juice packing box and his later series of labels of Campbell's soup cans). In pop art, we can detect the theme of "authentification" again: by painting a familiar, commercial object, Warhol seems to authenticate our experience with it. But another theme of Warhol's work is the impact of mass production and its impersonalization, and von Hagens picks up this theme very deftly. His factory in Dalian, China, mass produces most of the plastinates used in his exhibits. Furthermore, inviting comparison to an art exhibition, he presents his subjects in a variety of "life-like poses," suggesting living humans, most posed as athletes.

Von Hagens also understands how to reference artworks and cultural symbols of the past. The Horse and Rider combination (also seen in the *Casino Royale* exhibit), for example, is suggestive of the preserved remains of Roy Rogers's horse, Trigger. Trigger was stuffed after he died in 1965 and was displayed, for many years, along with Rogers's preserved dog Bullet and an admittedly artificial Rogers in the old (now closed) Roy Rogers and Dale Evans Museum in Branson, Missouri. Trigger was sold at auction in 2010 for an estimated $266,000. Trigger's executors point toward an interesting question: the position of plastinated human bodies in the commercial environment. Is a taxidermically preserved Trigger a saleable object? How about a polymer-infused human body? Is there an intrinsically different ethical question in charging admission to an exhibition versus disposing of an inorganically preserved human corpse by selling it? Are they art? Are they human remains? Are they educational scientific displays?

Von Hagens has set forth his views on the plastinated corpses' capacity as art:

> From *my perspective* plastinated specimens are not works of art, because they have been *created for the sole purpose of sharing insights into human anatomy* . . . a plastinator is at most a skilled laborer in the field of art, but not an artist as such. It would be completely misguided to refer to posed specimens as works of art, because (modern) art is a term subject to interpretation; as a result, everyone projects his or her own personal understanding of art and morality onto the motivation underlying *my efforts.*
> (emphasis added)

Von Hagens goes on to discuss his own views of why it is critical that his specimens appear "real" rather than as an artistic expression of reality:

> The *realism* of the specimens contributes greatly to the fascination and power of the exhibition. Particularly in today's media-oriented world, a world in which we increasingly obtain our information indirectly, people have retained a keen sense for the fact that a copy has always been intellectually "regurgitated,"

and as such is always an interpretation. In this respect the Body Worlds exhibition satisfies a tremendous human need for *unadulterated authenticity*.[3] (emphasis added)

Both of these quotations from von Hagens also illustrate, as does the *Casino Royale* press release, his fixation on himself. One of his admirers suggests, "His image cultivation is mainly a means of promoting *Body Worlds*. An inherent danger of such self-promotion is that the discussion may quickly focus on the producer of the exhibition rather than the exhibition itself."[4] It has also been reasonably suggested that "running throughout all von Hagens' endeavors there is . . . an attempt to escape from the reality of death, but giving the impression that these cadavers are continuing to exist in much the same way as when they were alive."[5]

The first time I saw the exhibit was at the Boston Museum of Science (where it was presented as an educational science exhibit in 2006). I was invited to comment on the ethics of the exhibit at a public forum sponsored by the museum. I titled my remarks "Dead Bodies, Dignity, and Property" and focused on four issues: the body as property, gender modeling in the *Body Worlds* display, consent to be plastinated, and the creation of what I called a "post-mortem advocate" who could exercise the right to withdraw (i.e., to bury or cremate) a deceased when the advocate determined that the deceased would no longer approve of the display of his or her remains. I drew heavily on the excellent article by Y. Michael Barilan that had just been published.[6] I concluded with four suggested "new rules" for plastination: that informed consent is necessary but not sufficient, that a postmortem advocate position should be created, that custody of a corpse should be distinguished from ownership of a corpse, and that an endpoint to the plastination project should be defined. (I suggested the project should end when it became impossible— or at least very difficult—to distinguish the "real" human plastinate from an entirely artificial duplicate.)

Closely related to the question of goals and ends is the question of categories of display. Much discussion has centered on whether the plastinates are entertainment or education (they are certainly both), and somewhat less has considered whether they are art. The art question is, I think, more significant and merits further exploration. My primary interest in this question is derived from my own work on bioterrorism. In that sphere, it has been difficult to distinguish bioart from bioscience or bioterrorism, and not just for the FBI. As I have argued, often the distinction is primarily a matter of identifying the researcher/creator's intent. Bioartist Steve Kurtz, for example, creates live art, using such creatures as deformed flies and colorful bacteria to create confusion and what he calls "fuzzy biological sabotage." One of the bacteria he uses, *Serratia marcescens*, has also been used the same way he planned to use it (before he was arrested by the FBI) by biological weapons designers to test the dispersal range of a biological weapon—which could also be used to sow terror. The same bacteria is used the same way; in one use the intent is to criticize the government through art, in the other use the intent is to test a weapon.[7]

Plastinated bodies in lifelike poses echo some examples of performance art—both because they are human bodies and because they are frozen in time. The performance artist they most mirror is Marina Abramovic, whose recent show at New York's Museum of Modern Art (MoMA), "The Artist is Present," reminded me of *Body Worlds*. Many of her performances have involved her being totally still in various (lifelike?) poses for long periods of time, approaching the end of human endurance—a stillness that is possible (for most of us at least) only in death. Her performance "Nude with Skeleton" (2002–5), in which her naked body is covered only with a human skeleton, was recreated by her students for more than six hundred hours during the MoMA exhibit. *Body Worlds*, of course, positions itself somewhere in the transition between live body and skeleton—freezing the muscles and organs in time, but both require us to ponder death as the inevitable consequence of having a body in the first place.

Some of Abramovic's performances, unlike *Body Worlds*, have been downright dangerous, pushing the boundary between life and death. In one, for example, she constructed a five-pointed star from wood shavings soaked in 100 liters of gasoline. She lit the star, walked around it, cut her hair and nails, threw them into the fire, and then lay down in the middle of the burning star. Her life was saved by a physician. As she recounts it: "I was supposed to stay there until it burned down, but as I was lying there the fire took up all the oxygen and I passed out. Nobody knew what was happening till a doctor in the audience noticed it and pulled me out. This was when I realized that the subject of my work should be the *limits* of the body."[8] Unlike the Abramovic performances, the *Body Worlds* exhibits pose no danger because there is no life in the plastinates.

A second performing artist to note, within this life-to-death spectrum, is the nonpracticing physician, Jack Kevorkian. Like von Hagens, Kevorkian is also fascinated by death, although for Kevorkian the fascination is more with what happens at the moment of death rather than after death. It may seem a stretch to view what Kevorkian did with his self-designed "suicide machine" as performance art, but there seems little doubt that his "final act"—videotaping his movement from assisted suicide to euthanasia by giving patient Thomas Youk a lethal injection—was meant to be just that. Kevorkian videotaped his performance and gave the tape to CBS's *60 Minutes* (which played it as a news/entertainment hybrid). He responded to Mike Wallace's question, "You were engaged in a political, medical, macabre publicity venture, right?" by answering, "Probably."[9]

But the "performance artist" who trumps these other two was Boston Red Sox baseball superstar Ted Williams. His death in 2002 occasioned much more controversy and discussion in Boston about corpses than did the *Body Worlds* exhibition in Boston in 2006. The reason was a dispute between his son and daughter over his remains. The daughter wanted them cremated (as his will had decreed), but his son, John Henry Williams, arranged for his body to be shipped to the Alcor Life Extension Foundation warehouse in Arizona. Ted Williams's body was to be cryopreserved (frozen in liquid nitrogen) for possible revival in the future when the damage to his body (caused by strokes and heart disease) could be repaired so he could continue living—or be born again.

Cryopreservation enthusiast Brian Wowk, not referring to von Hagens (but he might as well have been), early on suggested the problems his fellow cryopreservationists would have in selling their concept to the public:

> How often have we struggled with impressions that cryonics is a *sacrilegious, ghoulish, or Frankenstein-like* practice when we try to explain the concept? How often have we had the impossible task of trying to overcome the notion that cryonics entails supernatural resurrection when we try to explain its scientific foundations?[10]

Wowk thought that the reason for such difficulty was in the problem of explaining death and distinguishing the thawing process from Frankenstein's reanimation project. He thought the best way to get past this difficulty was to consider the frozen remains of real human bodies as not dead at all but existing, rather, in a state of "cryonic suspension" waiting to be thawed. In this regard he seemed to consider them in the category of frozen human sperm and embryos—in a frozen state, in which activity is suspended, but life continues.[11]

Wowk was certainly correct to observe that the cryopreservation process appears to be at least bizarre and even "ghoulish," and the same can be said for plastination. On the other hand, the goals of "freezing" corpses in time are radically different from those of freezing them in anticipation of a future thawing. Von Hagens wants to preserve the physical body (or at least major parts of it) as is in order to display it; the cryopreservationists also want to preserve the physical body, but their goal, in hopeful audacity, is to (re)animate it in the future, so it can live again. The weirdness continues with the choices offered by the cryopreservationists: you can have your whole body frozen, or (for considerably less money) you can have just your head frozen (the idea is that by the time you are defrosted, science will have found a way to "clone" your body and attach the new body to your preserved head). Of course, many obstacles have to be removed before one's frozen body is thawed, but the true believers, including Alcor's president at the time that his facility hosted the freezing of Ted Williams, believe that nanotechnology holds the key to solving these problems: the disease that caused your death has to be cured, the aging process has to be reversed, and the damage of freezing and storage has to be repaired.[12] The homage to the positivist intellectual tradition is awe-inspiring.

Unlike Vesalius's men and the plastinates, cryopreserved bodies are stored upside down (to preserve the heads in case of a liquid nitrogen shortage), and they are in stainless steel containers and thus are not viewable or on display. On the other hand, like the plastinates, they are "real human" bodies. Like many of the plastinates, the body of Ted Williams is the body of an athlete, although he is not preserved in the "real-life pose" of a hitter. Sportswriter Dan Shaughnessy expressed what may be a consensus view:

Ted on ice. Freeze-dried Ted. The Frozen Splinter. Could this be any worse? Stripped of any chance for a dignified burial or cremation, the body of the greatest hitter who ever lived rests in a cryonic warehouse in Scottsdale, Arizona. There will be no funeral, no memorial service. Instead, Williams' remains will be housed in very cold storage, with the wacky hope that someday he'll be back among the living. If this is what Ted wanted, he never told anyone.[13]

Not exactly death with dignity. Ted's plaque at the National Baseball Hall of Fame reads, in part: "Theodore Samuel Williams had only one goal in life: to walk down the street and have people say, 'There goes the greatest hitter who ever lived.'" Would either his reputation or the visitor's experience be enhanced by having at the museum an Alcor vat with his remains in it, his plastinated remains (in a hitting position), or a wax figure of his body (like Babe Ruth's)? Or is it better to have movies of his real-life hitting on display? Which immortality would Ted want? The baseball immortality that he achieved by his hitting or the bodily immortality that his son seeks for him?

Von Hagens will, of course, die; but he will achieve a sort of immortality through his plastinates. They're not "real bodies" and they won't live on, any more than James Bond will or Ted Williams seems likely to. Nonetheless, their "lifelike" presence will continue to evoke a frozen time somewhere between life and death that can entice us not to deny death itself but rather to consider alternatives to burial and cremation.

The good news about Americans is not that we don't deny death—we do—but that we're not entirely insane on the subject. Fewer than one thousand of us seriously plan to have our bodies cryopreserved or plastinated. Alkaline hydrolysis anyone?

CHAPTER THREE

More Wondrous and More Worthy to Behold

The Future of Public Anatomy

GEOFFREY REES, PHD

Since knowledge of the human body has enormous use, in no way should the anatomical teaching be neglected. It is always a worthy matter for man to behold the nature of things and not to despise the consideration of this wondrous work of the world, which was so skillfully created that it reminds us about God and His Will, as if we were watching a theatre. But it is most befitting and useful for us to see in ourselves the series of parts, figures, connections, powers and duties. It is said that there is an oracle "know thyself" which, although it advises many things, may here be taken to mean that we should earnestly behold those things which in ourselves are worthy of admiration and are the sources of most actions in life. And because men were created for the sake of wisdom and justice, and true wisdom is knowledge of God and consideration of nature, it should be acknowledged that anatomical teaching in which the causes of many actions and changes in us are seen, should be learned.

PHILIPP MELANCHTHON, "ORATION IN PRAISE OF ANATOMY," 1550

To the maxim "everything old is new again," the exhibitions of plastinated human bodies traveling the globe are no exception. Although the scale of these exhibits is unprecedented, as is their popularity, in spirit they hearken back to a prior practice of European humanist education. Just as the Renaissance was characterized by a renewed emphasis on all things human, including the human body, at the same time a central tenet of the Reformation—encapsulated in the doctrine of *sola scriptura*—was that all persons should read scripture directly and not depend on others to interpret it for them. As early as the middle of the sixteenth century these two strands converged in a focus on human anatomy. This convergence is exemplified in the epigraph from Philipp Melanchthon, one of the most influential institutionalizers of Reformation educational practices. In his "Oration in Praise of Anatomy" Melanchthon analogizes the human body to a theatrical performance of God's word, so that the body becomes a kind of text open to direct interpretation, and at the same time he invokes the classical injunction of the oracle at Delphi to "know thyself" as a human being. (This is the same oracle, incidentally, that Socrates in Plato's *Apology* says inspired his pursuit of knowledge of what makes excellence in a human being and a citizen.) The result was that all human beings were thought to have a compelling interest in learning directly for themselves about human anatomy. Its study became a path to knowledge of God and self.

The interest in anatomy exemplified in Melanchthon's oration reached its dramatic peak during the seventeenth century, when public dissections of human cadavers were celebrated as a valuable civic and philosophical entertainment. Many European cities built "theaters" expressly for the purpose. In principle, if not exactly in practice, these performances of dissection promoted the ideal that knowledge of the human body was of interest to all human beings, that it was too important to restrict to medical practitioners. So the finest seats in the theaters were reserved for the most important citizens of the community, while medical students were relegated to the galleries. Intended as much to entertain as to edify, the anatomic theaters realized practically the vision articulated by Melanchthon.

Although anatomy remained open in varying degrees to public study beyond the heyday of the anatomic theaters, by the middle of the nineteenth century the importance of anatomy to the possibility of self-knowledge for all persons gave way to the consolidation of medicine as a modern professional discipline. Part of the mystique and authority of the tradition of medical education for the past 150 years, indeed a rite of passage that traditionally makes doctors different from ordinary persons, has been the anatomy lab of first-year medical students. The anatomy lab has been closely restricted to first-year medical students, so that to become a doctor has been to gain privileged access to the interior of the human body. The anatomy lab of modern medical schools, extending the metaphor of the body as a "text," institutionalized a reversal of the Reformation spirit of common inquiry and ministry that urged persons to learn to read the body for themselves. Doctors, in the process, became a new sort of "priesthood" with exclusive authority to interpret the human body.

Considered in such a historical perspective, the great novelty of the plastination exhibitions seems to be a function of their return of the study of anatomy to the public stage and not of any particular detail of the exhibitions themselves. In other words, these exhibitions are so novel because they create a global population of first-year medical students able to purchase their education in anatomy for the cost of a day's admission to the museum. In this sense these exhibitions truly are extraordinary. As a means of edification and entertainment, however, they break little, if any, new ground.

With regard to the intention to edify, throughout the history of medicine the death of some persons has been an occasion for the education of other persons. The ensuing cumulative advances in medicine have enabled progressively more effective medical care for everyone. The plastination exhibitions simply enlarge the community of persons who can learn directly from the mortal bodies of their predecessors in death. While it is certainly true that the availability of accurate and comprehensive displays of human anatomy to ordinary citizens marks a return to an ideal of general education that is in tension with the modern history of medical training, it makes little sense to object to that return unless one is going to object to the teaching of anatomy in medical schools. Alternatively, one would be willing to argue that somehow admission to a medical school entitles a person to special knowledge about the human body to which ordinary citizens are not entitled or deserving, as if ordinary persons somehow have no right to knowledge about their own bodies.

With regard to the intention to entertain, the most distinctive features of the plastination exhibitions are their scale and global reach, since the display of preserved human remains in various forms has been a feature of museum collections for several centuries. Objections to the plastination exhibitions consequently also make little sense unless one is prepared to seek to shut down the whole realm of related traditions of exhibition. Almost every major museum of art, for example, includes a prized collection of Egyptian mummies. And almost every major museum of natural history or science exhibits a variety of bodily artifacts and would jump at the chance to add a bog man or a similarly preserved human corpse to its collection, if it doesn't already possess one. Likewise, small museums of medicine throughout Europe and the United States have quietly and uncontroversially been offering displays of preserved human curiosities since the end of the eighteenth century. The most relevant example is the Musée Fragonard d'Alfort in France; for more than two hundred years, this museum has been displaying the kinds of spectacular flayed corpses that Gunther Von Hagens has since reproduced and made so familiar to audiences around the world. Even Von Hagens's most famous tableau—"Rearing Horse and Rider"—is a copy of a tableau by Fragonard, based in turn on an image of "The Four Horsemen of the Apocalypse" as famously depicted by Albrecht Dürer.

The most remarkable accomplishment of the plastination exhibitions, then, is less their innovation and more their marketing, less the novelty of the forms they display and more their ability to present to a global audience a longstanding albeit somewhat esoteric European tradition. With regard to the end of the self-knowledge that inspired that tradition, however, the plastination exhibitions offer only limited improvement and in some ways even fall short. Consider, for example, that the plastination exhibitions are very static affairs that keep their audiences at a safe distance from any of the actual work of preparation of the bodies on display. By contrast, the anatomic theaters were an open process. Beginning, admittedly gruesomely, with the public execution of the person whose body was to be dissected, the theaters allowed the audience to observe directly the process. In doing so they did more than just present the anatomy of the body; they invited the audience into the body, into the process of exploration of the body. Even the order of dissection of the body in the anatomic theaters was dynamic and educational, since it necessarily began with the viscera, which decompose the quickest. The theaters taught not only about anatomy but also about the process of decay of the body, about the relative impermanences of the body, whereas the plastination exhibitions present a false impression of the body as entirely fixed. The theaters even included interludes of live music and reading aloud from classic medical texts.

Whereas the anatomic theaters really did provide vivid explorations of the human body and especially of the death of the body according to the best understanding of their time, today's plastination exhibitions promote a definition of death that is no longer medically current. As much as they do provide comprehensive displays of human anatomic structures, they are also vast collections of memento mori grounded in the European traditions of *vanitas* and *nature morte* painting. In remaining true to this tradition they effectively purvey at the beginning of the twenty-first century an understanding of death that was state of the art in the sixteenth and seventeenth centuries. They are therefore best regarded as the final word in a long sentence of medical education and entertainment just at a moment when a new paragraph has already begun.

The first sentence in that new paragraph was written in 1968, when the Harvard Ad Hoc Committee on Brain Death published its famous report. By 1980 the Uniform Determination of Death Act, which establishes brain dead as legally dead, became the law of the land. Since then, similar measures have been adopted throughout much of the world. Given that most of the ethical questions asked about the plastination exhibitions concern the uses of dead bodies for purposes of public education and entertainment, it is only logical also to ask those questions about the uses of brain-dead bodies, especially because the potential of brain-dead bodies to advance the goals of these exhibitions is exponentially greater than the current use of plastic corpses. Given, furthermore, that the great success of these exhibitions is a de facto argument for the permissible use of dead bodies for these purposes, it is only logical also to seek to take advantage of the more current scientific understanding of death as brain death to develop those purposes. Whether the goal is to astonish audiences with the complexity and the beauty of the human body or to teach them as accurately as possible about the structure and functioning of the body, brain-dead bodies provide an amazing opportunity to advance and surpass the pursuit of self-knowledge idealized by the anatomy theaters far beyond the accomplishments of the exhibitions of plastinated bodies now garnering so much attention.

What will be some of the advantages of exhibitions of brain-dead bodies, in contrast to the current generation of plastination exhibits? For the purposes of education about the body, the potential of brain-dead bodies to teach the general public is almost without limit. The most remarkable promise of their use is that they will finally bridge the divide between anatomy and physiology that has haunted medical inquiry from its beginnings. Throughout the history of medicine, the opportunities for the study of physiology have always trailed the opportunities for the study of anatomy. This is because anatomy, concerned with the static structures of the human body, is profitably studied only when organic vitality ceases. Physiology, by contrast, describes the dynamic functioning of bodily structures and, as such, cannot be studied once the vital functions cease. Before development of the technologies that enable the maintenance of bodily functions after brain death, it was never possible to correlate entirely structures and functions. The time, however, has now arrived when carefully managed and presented brain-dead bodies will allow the widest possible community of learners to observe the human body in all its functional purity, to observe anatomy and physiology all at once, to learn not just how the body *looks* from the inside but how it *works* from the inside. Instead of using static, flayed plastic bodies, designers of future bodies exhibits will be able to flay the brain-dead body so that viewers can observe all of its parts functioning in relation to each other. Instead of looking at a plastic kidney hanging like a dried bean on a wall, a liver on a shelf next to it, a heart on a pedestal, learners will be able to see a brightly perfused kidney making urine, a liver producing gall, a heart pumping blood, all carefully isolated, and then relate all those isolated physiological demonstrations in turn to a whole body of which they form only the parts.

An additional educational benefit of the use of brain-dead bodies is the advancement in medical research and training that they will facilitate. Plastination is often touted as a great innovation, but it is better characterized as a perfection of an ancient human practice just at a time when much more remarkable and truly innovative technological possibilities are at hand. While the basic means for exhibition of brain-dead bodies already exist in the forms of ventilators and pressors and other machines and medicines standard in hospital settings, all those means are presently limited in application to the treatment of brain-dead bodies as nothing more than husks to be harvested for transplantation. The development of technologies to enable displays of brain-dead bodies also promises to contribute to better clinical outcomes in hospitals by opening a much richer field of exploration. Innovations in surgical technique and surgical safety, for example, will be greatly enhanced. At the same time, accurate and complete displays of the functioning human body will enable the general public to understand more directly the anatomic and physiological basis of the treatments and procedures they will likely have to choose among for themselves or family members during their lives. (Already all sorts of educational videos of surgeries are publicly available for persons to review prior to undergoing procedures.)

As a form of entertainment, the possibilities for display of brain-dead bodies and their isolated components are only limited by the ingenuity and imagination of the bioengineers who will work with the bodies. Certainly it will be possible to develop exhibitions that are much more interactive and engaging than the static body exhibits. Special exhibitions might allow visitors to direct and manipulate the bodily functions. Similarly, interactive exhibitions might allow visitors to direct an endoscope throughout a body or to cause a muscle to contract or expand on demand. Perhaps, working with genetic engineers, the designers will even be able to scale the exhibitions so that visitors will be able literally to journey into the interior of the body. And all of these possibilities will only build on a trend among artists around the world currently exploring creative uses of the body and bodily materials.

Just as some persons have raised objections to the plastination exhibitions, objections will likely arise to the improved generation of exhibitions of brain-dead bodies. But most of the possible objections can be anticipated and effectively addressed. Some will protest that these exhibitions are disgusting or repugnant, as if the "yuck" factor were enough to argue against them. But many persons have a similar response to reptiles, cephalopods, and other organisms that are displayed in aquariums, zoos, and museums of science, and no one accepts that such a response actually constitutes a reasonable argument against those displays. Furthermore, images of vitally functioning human bodies already permeate the airwaves, so that any one who is objecting on these grounds probably hasn't bothered to watch the Discovery Channel or even public television, never mind visit the website YouTube. Already in 1978 on PBS the British neurologist and theatrical impresario Jonathan Miller reintroduced anatomy to a public stage when he broadcast an autopsy on his series, *The Body in Question*. Since then, television shows and websites full of images of surgeries and transplants and bodily exploration have multiplied exponentially.

Furthermore, one ought to question why any person would consider human anatomy and physiology disgusting or repugnant in the first place. Persons who object on these grounds are more likely expressing their own discomfort with the truth of their bodies than they are expressing any justifiable moral objection to the use of brain-dead bodies to present anatomy and physiology to the general public. As with any other popular media, the most reasonable response by persons who find them disgusting or repugnant is not to watch, but not to censor others from watching. Most important, persons who raise this objection fail to recognize that an initial response of fascinated repugnance is a draw to such exhibitions and that this response joins the aims of education and entertainment precisely because it creates an opportunity to explore openly and creatively one's own thoughts and emotions about one's embodied reality. Repugnance is best understood not as an end in itself but as an invitation to reflection about the reality to which one is responding.

Other persons might raise questions about the possibility of informed consent by the donors of their brain-dead bodies, such as have been raised about the plastination exhibitions. Here, too, the objections are misplaced, since the precedent for donation of organs from brain-dead bodies is actually much stronger and better established than that for donation of one's whole body to a plastination exhibit. Government-sponsored agencies around the country and the globe already currently solicit persons to donate their organs when brain death occurs. In most states in the United States of America, the question is joined to such a mundane activity as renewal of one's driver's license. In other countries consent for organ donation is even presumed unless one explicitly opts out. In all countries that have acknowledged brain death as legal death, a host of safeguards are in place to prevent conflicts of interest, to ensure that the use of one's brain-dead body is kept strictly apart from the care of one's body before brain death, so that no one is intentionally killed for the purposes of use of a brain-dead body. What, then, is the difference between donation of one's heart, liver, kidneys, corneas, and other tissues for transplantation and experimentation and donation of one's whole body or parts thereof for exhibition in a museum if one so desires? Both are acts of generous donation that serve valuable public goods.

Still others will complain that the display of brain-dead bodies and related body parts somehow insults the dignity or sanctity of the body. But it is difficult to explain the difference in principle between such displays of brain-dead bodies and their uses for transplantation and research. And, likewise, it's hard to say how such displays in museums would be any less dignified, for example, than the displays already beaming into people's homes and available on the Internet. Or how they somehow would insult or undermine the "sanctity" of those bodies. Here it is relevant to point out that the display of relics of human bodies has always been a form of respectful remembrance of the dead, a medium of intensification of sanctification of the body. Religious traditions around the world for millennia have been cherishing bodily relics of the revered among them, from bones, hair, and teeth to whole bodies. Unless one is prepared to argue that all these traditions are somehow affronting the dignity and sanctity of the dead body, there is no inherent reason to regard the uses of brain-dead bodies as necessarily less reverent. Perhaps the use of brain-dead bodies for these exhibits will even help inaugurate the beginning of a new era of sainthood in which the brain-dead bodies of the most famous or enlightened or revered leaders will be preserved as sanctified objects of public worship for religious veneration (e.g., the body of the pope) or secular celebration, just as the body of Lenin was preserved at the Kremlin for so many years.

Once bioengineers have figured out how to maintain brain-dead bodies, it is only a matter of time, some persons might worry, before brain-dead bodies will be used in all sorts of additional economically productive ways, whether for the culture of tissues, the manufacture of blood products, or even as zombie laborers. But the reality is that the use of deceased human bodies for commercial ends is already a huge business, so that much of this worry is belated at best. Indeed, such use is actually an age-old business. Relics in religious traditions have also always been valuable economic resources, just as mummies today are often the most popular draws to museums of art and natural history. Furthermore, as with other slippery slope arguments, people who worry about this prospect make the mistake of assuming in advance that everyone agrees already with their assessment that certain ends are objectionable, when in fact such agreement doesn't exist; they also make the mistake of assuming that no means will exist to prevent further steps once a first step has been taken. Both assumptions are false. Just as some persons welcome the prospect of the use of their dead bodies in medical schools and the body exhibitions, some persons might welcome the prospect of additional productive uses of their brain-dead bodies after their death rather than allowing their bodies go to waste in the earth. Furthermore, there is no reason to expect that additional future uses of brain-dead bodies can't be regulated at least as effectively as the currently permissible uses of brain-dead bodies.

Finally, it is important to acknowledge that there are significant practical concerns that will have to be addressed, most of all the possibility that the use of brain-dead bodies poses a public health hazard because these bodies cannot be antiseptically maintained and will become vectors for the development and spread of infectious agents. As with the risks associated with other valuable public works, however, there is no reason to expect that they cannot be managed responsibly. The use of hoods and similar technologies already standard in laboratories promises an immediate way forward. Even more promising, looking further ahead, is the possibility of advances in the plastination process itself, more in the mode of cyborg technologies. Although the current process tranforms all of the bodily tissues into a static antiseptic compound, a worthy goal will be to develop a version of the plastination process that similarly transforms only some tissues while allowing others to continue to function, creating dynamic hybrid vital bodies, more like cyborgs than mummies, that will be easier to manage, be less prone to decay, and minimize risks to public health.

Long ago the oracle at Delphi commanded every human being to seek knowledge of self. There is no reason to believe that the oracle exempted death itself from that injunction. Although the fact of death remains a constant of human life, the pursuit of knowledge of self involves acceptance that understanding of death is not static but develops with the development of biomedical science. The discovery of brain death not only is such an advance in human self-knowledge on its own but also breathes new life into the spirit of inquiry that once animated the productions of the anatomic theaters. The use of brain-dead bodies will enable extraordinary possibilities to educate and entertain the general public with animated displays of the beauty and complexity of the human body. Failure to adapt the body exhibitions to the best and most current understanding of death will therefore be to cling to an outmoded tradition of death and in the process to deny persons rightful discovery of the full truth of their bodies according to the best scientific knowledge and standards of their times.

CHAPTER FOUR

Resisting the Allure of the Lifelike Dead

CHRISTINE MONTROSS, MD

W HEN WE WANT TO BE, we humans are easily tricked. In psychiatry, we talk about denial; in theater we deal in suspension of disbelief. Both are at play in the current widespread public enthusiasm for touring exhibits of dead people.

As we know from both science and drama, it doesn't take much to convince us. However implausible the explanation or prop, we're willingly (mis)led: the wife who is told by her husband this really will be the final time he uses drugs; the audience that gasps and sits in silence after hearing an offstage gunshot.

In the museum exhibitions of plastinated bodies, the props are sometimes simple and spare—sets of bright, wide-open glass eyes, for instance, which look out from precisely dissected facial muscles. But the staging mechanisms are far less subtle. Almost without exception, the bodies are posed in action. Here a dissected goalie lunges in front of a goal to block a kick! Here a cross-sectioned rider straddles a flayed horse that rears up on his hind legs, even in death! At every turn, whether by bright eyes or lunging athletic maneuvers, the message conveyed to viewers is a Wizard of Oz–like deception: You needn't look behind the curtain because the dead don't do *this*. These bodies can't really be dead—they look so confoundingly alive.

I first attended one of the touring exhibits shortly after I had finished my own dissection of a cadaver as a medical student in gross anatomy lab. My experience of dissection had been filled with ambivalence—it was as fascinating as it was grueling; the proximity to death left me simultaneously troubled and deeply inspired.

During our semester dissecting cadavers, the anatomy faculty had given my classmates and me permission to invite friends and family members to visit the lab with us on occasion. I issued this invitation frequently for people to see the work that I had been doing so that they could see firsthand what a heart or a lung looked like. The responses I received were highly varied. My partner Deborah, whose mother had had breast cancer, came on one of the semester's first days to see a dissected breast in which a cancerous tumor had been found. She stood back from the bodies, hands held at her sides. A month later, a philosopher friend who writes at length about issues of personhood spent nearly an hour with me in the lab, holding various body parts in his gloved hands, asking questions, taking a metal probe from the drawer and poking around. His wife made it as far as the entrance door of the lab, took one look at the body bags, and headed for the exit. When my parents visited me, I offered them a trip to the lab. My father opted to come with us to the medical school but wanted no part of viewing the cadaver. My mother, matter-of-fact and not at all squeamish, entered the lab with me and, with nonchalance, promptly stated that the bodies looked a lot like turkey carcasses.

Even for those of us who had explicitly chosen careers in medicine, our days in the anatomy lab were often punctuated by aversion, dread, or even disgust. It was this understanding of the profound emotional charge of viewing dissected cadavers that I carried with me into the *Body Worlds* exhibit at Chicago's Museum of Science and Industry, where I saw the dead and dissected bodies on display there and hordes of people thronging around them, seeming completely and utterly unfazed.

One elementary-age school group had designated the base of the flayed horse and rider as a meeting place; children sat beneath the raised hooves comparing stickers from the gift shop and peering into each other's handheld Nintendo screens. A gang of middle schoolers played a covert game of tag. Defying more than one teacher's orders to walk and not run, the pre-teens hid behind a skinned body leaning over a chess board and contemplating his next move, his spinal cord elegantly exposed. Other visitors circulated among the bodies on display: a giggling couple of twenty-somethings on a date, holding hands and nuzzling one another as they walked; a woman pointing to man's knee, ligaments exposed, saying to her husband, "right there, that's the one I had to have repaired."

The blasé indifference of the museumgoers who weave their way through plastinated corpses is, I think, entirely made possible by the way in which the bodies are animated. In contrast, what strikes one upon entering the anatomy lab is the motionlessness of the body bags on their stainless steel tables; the wrong hue of the cadavers' skin. There is an odd, disquieting surprise each time one takes a blade to a dead body and it does not flinch or cry out. We have no choice other than to be reminded, constantly, of what death looks like. The museum exhibits ask us not to confront our mortality—a predominant and unavoidable aspect of anatomic dissection—but rather to engage our sense of wonder about the body's abilities.

There is nothing new about the curiosity that we, the living, harbor for the bodies of the dead. We peer into coffins, stare at fatal accidents, and populate our prime time television hours with forensic investigations of gruesome deaths. Throughout history, examples abound of large crowds at public executions. During the seventeenth century, criminals could be sentenced not only to death but also to public dissection—a theoretically worse punishment that was meant to serve as a greater deterrent to crime than death alone. At one point, human dissection even became a high-society tourist trend: distinguished eighteenth-century visitors would be welcomed into candlelit anatomic theaters to witness a human dissection. Afterward, the group would go into an adjacent room to enjoy an extravagant feast.

For all of these viewings of the dead, the purpose—whether punitive or memorial—is to some extent to reinforce the dead-ness of the body. The purpose of the display is in the witnessing of incontrovertible stillness in a coffin or flames of cremation that eventually consume the flesh. The display signifies an irreversible passing that we rightly turn into a ritual. In doing so, we allow the mourning or the punishment to be made real. It is for this reason that in his essay "Tract," the wise poet and undertaker Thomas Lynch argues so vociferously for viewing the burial or cremation of a loved one:

> [You] should see [the burial] till the very end . . . Go to the hole in the ground. Stand over it. Look into it. Wonder. And be cold. But stay until it's over. Until it is done . . . If you opt for burning, stand and watch. If you cannot watch it, perhaps you should reconsider. Try to get a whiff of the goings on. Warm your hands to the fire. This might be a good time for a song. Bury the ashes, cinders, and bones. The bits of the box that did not burn. Put them in something. Mark the spot.[1]

It is easy in the museum exhibits of plastinated bodies to lose sight of the fact that the objects on display are not objects at all but bodies that just as easily might have been grieved over at the graveside or at the funeral pyre had they not decided in life (ostensibly of their own will) to instead be immortalized in a Sisyphus-like eternity. Is an afterlife spent guarding the soccer goal against the same immobile shot any less torturous than rolling a stone uphill?

If there is one thing that historical viewings of the dead have in common with the Bodies exhibits, it is reinforcing the message to the witnesses that whatever personhood existed in the body prior to death is now undeniably elsewhere. This fact is reassuring—it allows us to bury or burn our dead, knowing that the agony of the experience belongs only to those still living. If the afterlife is happening, it's not in the museum hall fielding goal shots. But it is also disconcerting: if the person we loved is not here within this body we have always known, then where?

The Bodies exhibits skirt that question by imposing unceasing, lifelike personas on the plastinated dead. There are also the appropriate, requisite questions of consent which the exhibits raise: If people around the world are signing up to be plastinated after death, why do so many of the bodies seem to be of Asian ethnicity? Why situate the dissection and plastination labs in China, a country with an appalling human rights record, inviting speculation that the bodies are former prisoners, executed criminals, or the indigent dead? How likely is it, really, that a woman of childbearing age signed up to be plastinated and died while pregnant? And how likely is it that that same woman's partner or family would have agreed to have the woman and her fetus plastinated and displayed, even if she had expressed a desire for such a thing while living?

Even if all proper requirements of consent are suitably met, there are additional troubling implications to the ways in which the exhibits depict the dead. Already there is a societal trend to mask the reality of death: our dead are embalmed and made up, as if preserving the body by preventing decay and disguising discoloration can prolong the life that is already irrevocably gone. But beyond that, our subconscious selves absorb constant messages from pop culture that death is reversible. Cartoon characters bounce back from death; soap opera leads are reincarnated or resurrected; video games allow numerous lives, enemy rampages, vicious combat, and then the same thing all over again for as many times as we like after a quick reboot. We begin to harbor a kind of magical thinking—maybe death isn't so final after all. This same conceit that death is surreal and alterable was invoked by the bodies I saw on display in Chicago.

Treating death like something we can cheat, escape, or conquer is more comfortable, but since it is also untrue, it is short-sighted. The danger in conceiving of death this way is that we will be robbed of the opportunities that a finite life provides. We see this repeatedly in medicine. Palliative and hospice care exist, in part, to ease suffering at the end of life and to allow people to have a dignified and comfortable death. But statistics show that referrals to hospice in America overwhelmingly come too late. Doctors hang on to terminally ill patients, attempting long-shot treatment after treatment and intervention after intervention, though the odds of meaningful recovery may be infinitesimal. As a result, patients are frequently referred for hospice care with only hours or days to live—too little time to allow for the exploration of what a patient's needs are at the end of life and to attempt to meet those needs.

Our Western aversion to death's inexorable nature is by no means universal. Buddhist monks are encouraged to meditate on the decay of human corpses. The Buddhist Sutra *The Foundation of Mindfulness* has in it a section entitled *The Nine Cemetery Contemplations* in which Buddhist monks are encouraged to view their own bodies on an inevitable continuum that ends with an unprettied vision of decomposition. "If a monk sees a body thrown into the charnel ground, and reduced to a skeleton, blood-besmeared and without flesh, held together by tendons," reads an English translation of one of the *Contemplations*, ". . . he then applies this perception to his own body thus: 'Verily, also my own body is of the same nature; such it will become and will not escape it.'" The Buddhist teacher Thich Nhat Hanh refers to this Sutra as the "meditation on the corpse" and suggests that even lay followers of meditation and mindfulness should

> meditate on . . . how the body bloats and turns violet, how it is eaten by worms until only bits of blood and flesh still cling to the bones, meditate up to the point where only white bones remain, which in turn are slowly worn away and turn into dust. Meditate like that, knowing that your own body will undergo the same process. Meditate on the corpse until you are calm and at peace . . . and a smile appears on your face. Thus, by overcoming revulsion and fear, life will be seen as infinitely precious.

In facing the inevitable reality of life as something with both a beginning and an undeniable end, the aim is to fully experience the present. How much can we cherish an existence that we tell ourselves has the capacity to be unending?

The acknowledgment of death allows us the opportunity to infuse our lives with greater appreciation and an appropriate urgency. There *is* a place for plastic models of the body. We *should* gape at the body's intricacies and peer deep into the structures of our complex and miraculous physical selves. But it is my view that those models should be models, and we should allow the dead to be dead.

When it comes to our human mortality, denial doesn't serve us well. In resisting the allure of the posed and lifelike dead, we preserve a delineation between animate and inanimate, dead and living. And in doing so, we save ourselves from a form of denial that has the capacity to rob our lives of precious meaning.

CHAPTER FIVE

Detachment Has Consequences

A Note of Caution from Medical Students' Experiences of Cadaver Dissection

FARR A. CURLIN, MD

K ATHERINE TREADWAY WROTE about her first experience as a medical intern responding to a "code blue."[1] In the aftermath of the organized but unsuccessful chaos that ensued, Dr. Treadway became uncomfortably aware of the everydayness of the process. She noticed how fluidly and casually the medical residents "returned to their rounds, picking up as though nothing had happened." In the face of this incongruence between the physicians' light-hearted demeanor and the awful gravity of the patient's death, Treadway was moved to pose and answer a question:

Where did we learn this detachment? For most of us, the first lessons came very early in medical school, when we were confronted with the dissection of a human body—conveniently called a cadaver, as though that made it something different from a person who had died.

A panoply of traveling museum exhibitions of chemically transformed and carefully dissected human cadavers now invite their viewers to learn a similar form of detachment. These exhibits introduce the layperson to an experience that, for the last century, has been reserved for health professionals. It is interesting to think about how such exhibits affect their viewers and how they might be restructured or reinterpreted, depending upon the particular emotional message that the exhibitors want to convey. Medical educators have long experience with the challenges of teaching (and learning) anatomy. Cautionary tales from medical school anatomy labs may be helpful for museum curators and museum goers both.

I have, for several years, given a lecture to the first-year medical students at the University of Chicago prior to their first anatomy class. In what follows, I will share what I've learned in preparing for and delivering those lectures and then consider the implications for the display of plastinated human bodies.

Viewing cadavers is not the same as dissecting them. Cadaver dissection, or "gross anatomy lab," is a defining experience of medical school. Over the course of several months, first-year medical students systematically disassemble a human body to gain familiarity with the human form that will be the object of their ministrations throughout their careers. Medical sociologist Fred Hafferty describes this initiatory rite of passage in detail in his book, *Into the Valley: Death and the Socialization of Medical Students.*[2]

Hafferty notes that cadaver dissection opens a divide between the laypersons that the students were and the physicians they are in the process of becoming. In the cloistered setting of the dissection lab, students act in ways that are not only impermissible for laypersons, but which most would find crude, repulsive, even criminal. They flay the skin off of their cadavers. They take a hacksaw to the cadavers and saw them into quadrants. They cleave the cadaver skulls in half. They wash stool out of intestines. They slice into, and examine, eyes, tongues, brains, hearts, and genitals. No wonder, Hafferty notes, the public finds itself

> distrustful of these practitioners who dare to enter the realm of
> the forbidden and the taboo. Although we are grateful for their
> ability to help us deal with our physical frailties, we also wonder
> how it is possible for them to do so, day in and day out, and
> still retain their own humanity. We applaud their fight against
> death, yet question whether their struggles have caused them to
> lose sight of what is important in life . . . We embrace, yet dread,
> our doctors' *detachment and their distance* . . . We are also

unsure when and if we have been *transformed from a patient into a tool for learning*. In short, we are fascinated by, yet deeply suspicious of, the transformation of lay persons into medical experts. (pp. 1–2, emphasis mine)

The public has good reason to be suspicious of young men and women whose professional formation requires experiences that are both alien to that public and morally disturbing. Does the dissection experience form these acolytes into the sorts of professionals to whom we want to reveal our deepest secrets and who we trust with our most treasured hopes? This may not be the best training to prepare physicians to decide whether a person is kept alive or allowed to die, though it undoubtedly gives doctors a unique perspective on judging whether someone has died.

Hafferty speculates that one of the functions of anatomy lab is to help teach physicians how to violate social norms that operate in every other social situation, a skill that will be necessary in clinical practice. The detachment that allows student physicians to cut up the dead may help practicing clinicians to put their hands and medical instruments in patients' various bodily orifices or to ask patients to confess their most shameful secrets and expose their nakedness in the most vulnerable positions.

None of these extraordinary powers and dangerous privileges would be granted for its own sake; in a nonmedical context, many of these behaviors would get an individual thrown into prison. They are granted only because they enable physicians to do what patients desperately need for them to do—namely, to attend to patients when they are sick, to diagnose their ailments, and to prescribe appropriate treatments. Doctors are given license to do the things they do—including mutilating the dead—only because they explicitly commit themselves to caring for the sick. Cadaver lab, then, is not merely an introduction to a macabre ritual. It is, instead, an invitation to begin to internalize a physician's most fundamental profession. Starting with that first dead body, the medical student will learn and discipline himself or herself to a set of practices that aim to care for the sick, relieve their suffering, mend their injuries, and, if possible, cure their diseases.

This training results in both intended and unintended consequences. Cadaver dissection, like the rest of medical training, is not only something students do but also something that is done to them. In the process of cutting up and gazing at human bodies, a student becomes someone he or she was not before. The moral challenge for medical students, throughout their years of training, is to navigate the tension that is made fully explicit in cadaver lab. That is the tension between medicine as a scientific, objective, technical practice in which the patient's body is a specimen and medicine as a moral, spiritual, subjective, vocational practice in which the embodied patient is recognized not as a specimen but as a fellow human, neighbor, and even friend.

The two worlds are not fully separable. Physicians in training inevitably experience ambivalence regarding which view of patients and their bodies should be emphasized at which times. They struggle to both embrace and resist the inevitable dehumanization. They try to hold on to their pre-training humanity, even as they realize the necessity of letting some parts of it go.

This tension has always run through the practice of medicine. More than one hundred years ago, Sir William Osler recommended a way for physicians to respond to the conflicting impulses. In a now famous graduation speech,[3] Osler suggested that doctors should seek a quality that he called *aequanimitas* (equanimity). Osler argued that aequanimitas is the defining characteristic of the good physician. Physicians with aequanimitas are composed, thoughtful, reasoned in their judgments, not easily disturbed, not overly attached. Like well-trained and disciplined soldiers under fire, these physicians lose no energy to distraction, fear, wavering, hurry, anxiety, and emotional responses to the suffering and deaths of those around them.

Aequanimitas is not easy to achieve. Physicians who err toward too much detachment and objective distance risk becoming jaded, callous, and unfeeling. Like soldiers who are skilled and effective mercenaries but who have lost the ability to discern the humanity of their enemies or their colleagues, such physicians endanger those they are meant to serve. On the other hand, physicians who err too much toward empathic connection with their patients might lose the objectivity and the detachment that allow them to make a dreaded diagnosis or perform a painful but necessary procedure.

The eminent medical sociologist Rene Fox further described the same tension using different terms. All healers are, at least in principle, animated by *concern* for the sick. Most prospective medical students come to medicine professing deep and usually genuine desire to help others. Yet to help those who are sick, physicians must encounter them in their frail humanity, and the humanity of patients is threatening. Fox suggests that physicians tend to objectify and detach from that humanity as a way of managing its threat while they go about doing their work of healing.[4] Medical training requires physicians to detach from the human particularities to learn objective principles, to learn the parts that make up the whole, to learn of the patient as specimen. Yet medical training also encourages physicians, at least explicitly, to maintain concern for the human particularities, to remember that patients enter our care not as generalizations or specimens but as fellow humans in need. Fox characterizes the balance that doctors seek as one of "detached concern."

Maintaining balance between detachment and concern is difficult. Detachment comes easily enough, beginning with the process of handling and gazing at human bodies as specimens in the cadaver lab. Of that experience Katherine Treadway writes:

> We learned to bury our fear of death in an avalanche of knowledge. We learned the trick of silencing the parts of our brain that didn't really want to be this close to death. And for good reason. We could not do what we do [as physicians]—take responsibility for the lives of our patients—if we were aware, minute to minute, of the true significance of what we were actually doing. We could not come into a code fully aware of the profound event taking place and still be able to do our job.

So we learned to put those feelings away. The question, of
course is how to avoid losing them altogether.

How, indeed, do physicians avoid losing altogether the concern that makes them human if the bulk of medical training seems to erode that concern?

Equanimity, or detached concern, are both means between, on one extreme, an emotional attachment that gets in the way of doing the things to a patient that are necessary to bring about health and, on the other extreme, an unfeeling detachment that fails to recognize and act in a way that accords with the humanity of the patient. The good physician will strike the right balance. Most physicians err on the side of detachment, rather than concern. Virtue, then, will consist of consciously striving to be more concerned, not because concern is more important than detachment but because we tend so toward detachment that only by strenuous effort and attention do we maintain proper concern.

This brings us full circle to the display of plastinated bodies. Body exhibits invite and indeed require the viewer to detach from the humanity and particularity of the body as the tangible remains of the individual who died. The modes of display all encourage the viewer to view the body as a specimen representative of a general type. This dynamic has predictable effects. It leads the viewer toward the sort of detachment that is achieved by medical students. Things that should inspire awe are turned into things casual and mundane. Things that would be sorrowed over are turned into things that are intended to pique and satisfy curiosity. In the context of medical training, this process is allowed only as the best means to achieve the end of allowing healers to learn how to heal. The knowledge is gained not to satisfy curiosity but to enable the student to participate in the noble moral activity that is the practice of medicine. Even then we find that the detachment that inevitably results threatens the humanity of the physician and thereby threatens his or her patients.

In recognition of this threat, medical educators have developed rituals and traditions to attempt to countervail the dehumanizing force of cadaver dissection. For example, in the gross anatomy course at the University of Chicago, I encourage students to develop small habits and rituals that help to remind them that they are dealing with something sacred. Simple rituals can help physicians reflect on and thereby take account of the significance of the life or death of any individual. Dr. Treadway wrote that she regularly whispers the words of a requiem Mass whenever she stands over a patient who has died, reciting "May choirs of angels greet thee at thy coming." Although Treadway discounted the religious connotations of those words, the words have power precisely because they point to hope that remains when the awfulness of death is set in some larger framework of meaning. For most that meaning comes from their religion.

In a secular, or at least nonsectarian, way, medical educators encourage students *collectively* to set the experience of dissection against some broader backdrop of meaning by holding cadaver memorial services. Many medical schools encourage students at the end of their gross anatomy experience to hold a service in which they commemorate those who gifted their bodies so that the students could become physicians, reflect on how they have been changed, for better or worse, and memorialize their experience so as to not take it for granted.

All of these efforts—rituals, memorials, and other reminders—are intended to help physicians maintain their own humanity without becoming overly detached from the humanity of their patients. To my knowledge there is no parallel experience with respect to displays of plastinated bodies. We may ask, therefore, whether body displays invite the public in subtle ways to treat the death of individuals, in the words of Treadway, "as though nothing had happened."

The experience of cadaver dissection alienates physicians from the public. The experience of gazing at a body exhibit might, in a parallel fashion, alienate the public from its own humanity by treating the human body as mere nature, mere tissues and organs. Exhibits of plastinated bodies are one expression of a view of human bodies as mere stuff. In *The Abolition of Man,* C. S. Lewis compares the modern scientific impulse to that of the magician—the desire

> to subdue reality to the wishes of men: the solution is a technique; and both [the magician and the scientist], in the practice of this technique, are ready to do things hitherto regarded as disgusting and impious—such as digging up and mutilating the dead . . . It is in Man's power to treat himself as a mere "natural object." The objection to his doing so does not lie in the fact that this point of view (like one's first day in a dissecting room [or one's first pass through the *Bodies Revealed* exhibit?]) is painful and shocking till we grow used to it. The pain and the shock are at most a warning and a symptom. The real objection is that if man chooses to treat himself as raw material, raw material he will be. [5]

The risk of exhibitions of plastinated bodies is not just that they will catalyze this process but that they will do so without the same recognition of moral risk that we bring to the similar process in the medical school anatomy lab. Thus, the point is not that people will disintegrate after seeing such exhibits. The fact that they get along just fine is precisely the concern. Getting over such squeamishness is crucial for a physician but not similarly crucial for everyone who can afford a ticket to a science museum. By conquering old reluctances and prejudices, we may be dismantling our few remaining defenses against treating humans as mere nature to be manipulated.

Gazing on dissected human bodies should never be done in a casual fashion. Medical educators recognize the danger of dismantling students' instinctive concern for the unique and particular humanity of the human bodies they encounter. The same cannot be said for those who make a living by charging admission fees to anyone who wants to see a flayed horse and its rider.

CHAPTER SIX

The History and Potential of Public Anatomy

CALLUM F. ROSS, PHD

MILLIONS OF PEOPLE have toured Gunther von Hagens's *Body Worlds* and the competing exhibits, demonstrating a powerful public interest in anatomy. This interest is not new. People have always been fascinated by human anatomy. They have satisfied this curiosity by attending public dissections, visiting anatomy museums, and, most recently, attending *Body Worlds* and *Bodies Revealed*. The potential benefits of anatomy museums are far-reaching and profound. Society would be better off if we had more of them.

The Public Interest in Anatomy

The nonphysician, nonscientist general public has long been interested in seeing human anatomy "in the flesh." In fourteenth-century Europe, there were opportunities for satisfying this curiosity. Universities regularly performed public dissections. These were performed primarily for the education of students and faculty but were occasionally opened to the public.

The first public dissection in Europe after the Dark Ages was allegedly performed in Bologna in 1315 by Mondino dei Liuzzi. Other universities or medical colleges soon followed suit. Permits were given for annual dissections in Venice (1368) and Languedoc (1376), for dissections of condemned criminals as circumstances permitted in Florence (1372), for triennial anatomic demonstrations at the University of Lerida (1391), and for annual dissections in the early fifteenth century in Bologna and Padua. In Scotland, the Surgeons and Barbers of Edinburgh were performing annual dissections of condemned criminals by the beginning of the sixteenth century, but apparently in private. The first public dissection in Edinburgh was performed in 1703 in the first Surgeon's Hall. The first authorized public dissections in England were performed on up to four bodies annually by the London Company of Barber Surgeons starting in 1540 and by the College of Physicians at Gonville Hall, Cambridge, starting in 1565. Public anatomy lectures targeted at medical students were given in the Jardin du Roi in Paris starting in 1643. Throughout the sixteenth century, annual dissections performed for medical students in Padua and Bologna in Italy could be attended by the public. Because they had to purchase or be given tickets, it is likely that only the wealthy elite could attend.

From 1638 through 1737 the public anatomy theater in Bologna was built and expanded to house public anatomy lessons that could extend to thirty separate lectures, open to the public for free, often as part of Carnival celebrations.[1] In Amsterdam the burgeoning popularity of public dissection from the late sixteenth through mid-seventeenth century necessitated moving the venue from smaller (Saint Ursula's Church) to larger (Saint Margaret's Church) facilities, then the construction of a public anatomy theater in Saint Anthonispoort in 1639, and its expansion in 1671. (Today, one can dine in the renovated Saint Anthonispoort in de Waag in Amsterdam: www.indewaag.nl/?English/.) After the French king decreed in 1673 that the Jardin du Roi rather than the Faculty of Medicine should have preferred access to the bodies of executed criminals in Paris, audiences at public dissections in the Jardin burgeoned into the hundreds. With the appointment of the young, attractive, and eloquent Joseph-Guichard Duverney, public dissections became fashionable, requiring in 1692 the construction of an anatomy amphitheater in the Jardin that could house audiences of up to six hundred. Interest in the public dissections by the Barber Company in London demanded the construction of new seating in the Barber Company's kitchen (where the dissections were performed) in 1567 and then the construction of the Barber Surgeon's Anatomical Theater in 1636. Public dissections continued in England and Scotland until the 1832 Anatomy Act confined dissection to licensed individuals and restricted cadaver access to medical schools. Thus, during the fourteenth through seventeenth centuries, public dissections, primarily of executed criminals, were well attended in Europe. These appear to have been mostly annual or, in some cases, quadrennial events.

While these dissections may have been intended to teach both moral lessons to the public on the consequences of violating the law and practical lessons to physicians in training on human anatomy, they also clearly functioned as venues for satisfying public interest in anatomy. In the middle eighteenth century, public anatomy demonstrations became unpopular in Europe as public ideas about death changed. It is probably no coincidence that, around the same time, collections of waxwork models were assembled. Originally constructed to teach anatomy to medical students, their accumulations in museums must have partly satisfied public interest in anatomy. Waxwork techniques were perfected in France and Italy in the early eighteenth century.

Guillaume Denoües brought his waxworks to England in 1720, and they became a staple of English anatomy museums until their demise in the mid-eighteenth century. At La Specola in Florence a natural history museum was opened to the public in 1775, making accessible a remarkable collection of detailed wax models based on dissections of human cadavers. These models were prepared in order to teach human anatomy without the use of cadavers, but their levels of detail and beauty give them artistic value. According to the La Specola website, "until the early years of the nineteenth century [La Specola] was the only scientific museum specifically created for the public." In eighteenth-century France, preserved human and nonhuman corpses were displayed at the Musée Fragonard d'Alfort, where one can still see a horse and rider (of the Apocalypse), human fetuses dancing a jig, a man brandishing a mandible (cf. Samson versus the Philistines), a wax-injected human head, and a dissected human arm. The museum was part of a veterinary school, and the human material was, presumably, presented to raise money for the museum.

In the United States from the seventeenth through nineteenth centuries executed criminals—primarily "blacks," Indians, indigents, or prostitutes—were dissected in public but apparently less frequently than in Europe. A Massachusetts law of 1647 limited dissections to one body every four years. The first U.S. anatomy theater, built in the 1760s in Philadelphia, was used for public dissection of Siamese twins but was primarily for the instruction of medical students. The comparative rarity of public dissection in the United States does not reflect a lack of interest in human anatomy, an interest that could be piqued in the right circumstances, perhaps given the right subject. For example, in 1836 P. T. Barnum hosted (in a saloon) a dissection of a black woman advertised (clearly fraudulently) to be the 161-year-old former nurse of George Washington; this spectacle attracted a crowd of fifteen hundred. Nevertheless, dissection as a form of public anatomy in America may have suffered from public outrage at body snatching and grave robbing. This outrage spurred a series of riots against medical colleges from the eighteenth through the nineteenth centuries. Instead of public dissection, public interest in anatomy in America was satisfied by popular lectures and anatomic museums and exhibitions. Many of these exhibits included waxworks and other models, in addition to actual anatomic specimens.

The Demise of Public Anatomy Museums

In the early to mid-eighteenth century, North American anatomy museums, with their waxworks and anatomic specimens, met a public desire for information on human anatomy and served an educational purpose. But by the late 1880s these museums had been marginalized by competition with state-sanctioned anatomy and physiology presentations in public schools and at hygiene fairs. To pay their bills, the challenged museums capitalized on the public desire for the sensational and lurid. Their emphasis changed from education to entertainment, with exhibits of torture, sexual anatomy, and the effects of venereal disease.

Public anatomy museums in nineteenth-century England declined in numbers for reasons that illustrate an important change in professional perspectives on dissection. In early eighteenth-century England, medical students were required to study anatomy, but only two medical schools offered anatomy courses. As a result, private anatomy schools sprung up that were not owned or operated by the hospital-based medical schools. The interested public, including artists and medical students, could attend the lectures and dissections if they could pay. Public anatomy museums, such as those of William and John Hunter and John Heaviside, were salons where the literati could attend lectures, view exhibits, and discuss matters of natural history. More prosaic public anatomy museums capitalized on public interest in anatomy by providing access to the same kinds of information as the medical profession. They presented anatomic specimens, detailed wax models, and even lectures on anatomy that were popular and well attended. However, in the mid- to late nineteenth century, professional anatomy instruction was centralized in and monopolized by hospital-centered medical schools, spelling the end of public anatomy museums.

The medical profession objected to public anatomy museums because many such museums had become the domain of quacks who sold false remedies for venereal disease. Kahn's of Piccadilly was begun as an anatomy exhibit that received glowing reviews from *The Lancet*. Kahn attracted the ire of the establishment, however, by profiting from association with purveyors of quack remedies for diseases, examples of which were on display in the collections. Although this was not the only, nor indeed the original, function of Kahn's anatomy museum, this association led the editor of *The Lancet* to call for its closure. Mobilizing the Society for the Suppression of Vice, *The Lancet* accused Kahn's and other anatomy museums of corrupting public morals. By displaying wax models of the genitalia of women and men, they fell foul of Victorian notions of public decency. They also had the poor taste to admit women to their exhibits, which was not excused even if they were midwives, nurses, and mothers. Under the Public Obscenities Act of 1860 public anatomy museums were closed down and their collections destroyed or purchased by medical schools. By the late 1870s all public anatomy museums and private anatomy schools in London were closed, anatomy instruction was monopolized by hospital-based medical schools, and public access to anatomic information was severely restricted. Thus, the demise of anatomy museums in England and America in the nineteenth century had different causes but similar effects: the severe curtailment of public access to anatomic knowledge. Public concern with the source of cadavers for dissection in both England and the United States and increasing professionalization of medical education further removed dissection from the public eye, reserving it for doctors in training.

There were still anatomy museums. They just lost their cachet. Even today, the Mütter Museum in the College of Physicians of Philadelphia, established in the 1850s, is open to the public. Initially founded to "educate future doctors about anatomy and human medical anomalies," today the museum also "serves as a valuable resource for educating and enlightening the public about our medical past and telling important stories about what it means to be human." The National Museum of Health and Medicine in Washington is open to the public and displays limited anatomic material but houses an embryological collection available to the public online. The Warren Anatomical Museum in Harvard's Countway Library of Medicine, open to the public free of charge, contains collections originally intended for the Harvard medical community. William Hunter's anatomic collections in the Hunterian Museum in Glasgow (www .hunterian.gla.ac.uk/collections/anatomy_index.shtml) are available for teaching and research but not for the general public. The Hunterian Museum at the Royal College of Surgeons in London presents some of John Hunter's collections, is open to the public for free, and can be visited online through a virtual tour (rcseng.ac.uk/museums/ galleries). These public offerings of human anatomy tend to be historical in nature and do not seek to attract the modern public audience with the aim of educating them about human anatomy.

Public Anatomy Education Today

The two leading plastination shows, *Body Worlds* and *Bodies Revealed*, market their exhibits as being educational. Clearly, however, their educational value is not fully explored and their primary role is entertainment. The absence of educational public anatomy exhibits is a disservice to the public. It could be remedied if the exhibitions were designed by professional educators, anatomists, museum curators, and psychologists. Museums should insist on this. The democratization of anatomic knowledge, like the democratization of all medical and scientific information, is the purview and responsibility of public museums.

On a more practical level, several examples of the educational benefits of anatomy exhibits can be envisaged. Clearly, powerful messages regarding health and disease can be sent using anatomic information on the effects of "lifestyle choices." *Body Worlds* includes preserved human lung material, blackened by carbon from smoking or pollution, and states that the "exhibitions aim to educate the public about the inner workings of the human body and show the effects of poor health, good health and lifestyle choices." Displays show the effects of lack of exercise and poor diet on cardiac anatomy and physiology, the effects of osteoporosis, the consequences of prostate and breast cancer and the importance of their early detection, and the benefits of good oral care.

Another important area where anatomy exhibits could benefit the public is the clear presentation of information on sex, growth, development, and reproduction. Rather than the moralistic diatribes centered around venereal diseases that characterized the nineteenth-century anatomy museums of England and America, a clear exposition of the facts of human sexuality and reproduction would be empowering for young adults and helpful to parents. A scientific presentation of the facts, based on anatomic specimens and models, would be a welcome addition to science museums in the United States.

Better public understanding of human anatomy also allows a scientific presentation of the evidence for human evolution, much of which comes from comparative anatomy of humans and other animals. Many human diseases can be traced, at least in part, to bad design as the result of evolution. This includes musculoskeletal disorders due to poor design of the back, the prevalence of hernias in males as a result of the developmental history of the testes, and temporomandibular joint dysfunction as a result of the evolutionary origins of the joint.

Why Real Bodies Instead of Models?

There are two advantages to the use of actual bodies, however prepared for display, rather than models. The first is a commercial consideration. Real bodies get people in the door. People come a long way and pay a lot of money to see real human bodies. The second advantage of using real cadavers is their accuracy. Wax models, such as those at La Specola, can be fantastically detailed, but they do contain inaccuracies. Real bodies differ from one another, and certain inaccuracies can be introduced during preparation, but these are of a different sort than those found in models.

The sources of the cadavers should be disclosed clearly to the public, the processes by which they are converted to displays should be transparent, and the disposal of the specimens should be as agreed to by the donors and their families. Although the use of real bodies has clear advantages, anatomy museums need not be entirely based on real human specimens. Real bodies may not be for everyone and, in the interests of being inclusive, models may be useful in some instances, such as educating the blind, the squeamish, or the young, some of whom may not wish to touch real cadavers. Moreover, models simplify, which is sometimes helpful, especially when presenting complicated concepts.

Dissection

Dissection is an important part of medical education. It is valuable because the acts of dissection and exploration are acts of learning. Through dissection, students learn where structures are located internally and relative to the body surface. They learn about connectivity of organs, nerves, blood vessels, muscles, bones, and lymphatics. These relationships and connections are necessary components of diagnosing disease and interpreting modern medical images.

Dissection seems less important for the general public. Nevertheless, it is not wrong or inappropriate for nonmedical people to want to see, or participate in, dissection. Currently, rules do not regulate who carries out dissections; they only regulate the institutions that can obtain cadavers, leaving the institutions with the responsibility of deciding who teaches and learns anatomy. A few universities allow undergraduate students to dissect human cadavers. Such courses could be offered outside the regular curriculum to other members of the public.

Public dissections might also meet an educational need. I can imagine a museum offering a series of displays illustrating the history of public dissection or an ongoing professional dissection of a cadaver, behind glass, with comment from professional anatomists. Such a well-regulated public dissection might be of immense value to interested students.

Conclusions

The public fascination with *Body Worlds* is merely the latest expression of an age-old public interest in anatomy. People's interest in seeing human anatomy in the flesh presents a powerful opportunity to provide educational opportunities for the general public. The history of anatomy museums is a mixture of attempts to meet educational objectives, make money, and dispense medical care. Modern anatomic museums need not be compromised in these ways; their exhibits could be professionally designed to meet very specific educational objectives and the content tailored to meet diverse needs and interests. The creation of these exhibits is, I believe, a responsibility that falls to museums, educators, and anatomists. *Body Worlds* has shown us that public anatomy can entertain. Museum curators should take seriously their responsibility to ensure that it also educates.

APPENDIX: MUSEUMS WITH SOME
HUMAN ANATOMY MATERIAL

Musée Fragonard d'Alfort
http://musee.vet-alfort.fr/

Museum of Zoology and Natural History at La Specola
www.museumsinflorence.com/musei/museum_of_natural_history.html

Mütter Museum in the College of Physicians of Philadelphia
www.collphyphil.org/mutter_hist.htm

The National Museum of Health and Medicine
www.nmhm.washingtondc.museum/exhibits/exhibits.html

University of Michigan Medical School Dissection Videos
http://anatomy.med.umich.edu/courseinfo/video_index.html

CHAPTER SEVEN

What Would Dr. William Hunter Think about *Bodies Revealed*?

LYNDA PAYNE, PHD, RN

E VERY CULTURE HAS CONVENTIONS about portraying and exhibiting the dead human body. Historically, what could be shown of the body—where it was displayed, how it was represented, and who got to see it—has always been a complex social issue.[1]

Dr. William Hunter (1718–1783) was the most famous anatomist and obstetrician in Europe and North America in the eighteenth century.[2] He ran the largest anatomy school in London, trained thousands of medical students—including 25 percent of the MDs in what became the United States—and had the most lavish and expensive anatomic museum in Britain. Hunter was born in 1718, the seventh of ten children of a Scottish Presbyterian farmer. Meant for the church, he disobeyed his father's wishes and apprenticed himself in 1734 to William Cullen, a local surgeon. In 1740, armed with introductions to the leading Scottish practitioners of surgery and man-midwifery (as obstetrics was known then) in London, Hunter departed for the capital and became, first, a student of William Smellie, the famous man-midwife, and then the chief dissector to James Douglas, a well-known surgeon.

With Douglas's wastrel of a son, Hunter attended a fashionable six-month course of anatomic lectures in Paris at the Jardin du Roi, or the King's Garden. There, however, Hunter was more impressed by the ready supply of corpses for anatomic instruction than by the course itself. In 1746, on his return to London, Hunter placed an advertisement in several newspapers announcing a course of medical lectures in his home, where the "Paris manner" of dissection would be used.[3] By this Hunter meant that each student, as in Paris, would get a body to dissect rather than watch a lecturer point out the parts on a body.

There were at least twenty-seven lecturers in anatomy working in London before William Hunter opened for business.[4] However, he quickly became the best known of them all. Besides instructing young men in the art and science of medicine, Hunter attracted noted Enlightenment figures such as Adam Smith, Edward Gibbon, and Tobias Smollett to his lectures. In 1762 Hunter was appointed Physician-in-Extraordinary to Queen Charlotte, George III's wife. He would deliver all fifteen of her children. By 1767, Hunter was a Fellow of the Royal Society. The same year he had a house, a museum, and an anatomic theater built on Great Windmill Street in Soho. It quickly became the most famous and prestigious anatomy school in the Western world.

Anyone could attend William Hunter's lectures as long as they had cash and were male. He offered instruction in a multitude of medical subjects other than anatomy, including operative surgery, the making of preparations, embalming, and midwifery.[5] So where did anatomy teachers such as Hunter—who literally required hundreds of corpses a year—obtain them? Only a few executed criminal bodies were earmarked by the British government for the use of surgical teaching; like other anatomy teachers, William Hunter had to partly rely on the services of body snatchers (also known as Resurrectionists, sack-em-up men, or ghouls) for the source of most of his corpses.[6]

Gangs of men and women broke into graveyards at night, having bribed any graveyard workers ahead of time to look the other way, and then dug up and carried off freshly buried bodies of adults and children. Some were sold locally to anatomy teachers; others were salted, pickled, and packed up for export. Both William Hunter and his brother, John, employed gangs of body snatchers and competed for the best human specimens in the ground and on the market. William Hunter drew up a list of regulations regarding the types of bodies he required—fresh, young, female, and pregnant were at the top of his list—and how exactly he wanted the bodies exhumed and carried to his school. He set prices at a guinea for an adult corpse and a shilling for the first foot and sixpence an inch after that for children's corpses.[7]

Given these life experiences, Hunter would have applauded the way *Bodies Revealed* marketed itself. He would not have been the least bit troubled by the shady sources of the bodies. In his era, under English and American law, no one could own a body, so stealing a corpse was only a minor felony. On the other hand, taking or selling any clothes on that stolen corpse was considered grand theft. The sentence for that was death by hanging.[8]

As artist-in-residence at the new Royal Academy of Art, Hunter would have liked the artistic presentation of plastinated corpses in *Bodies Revealed*. Similarly, he would have approved of opening the exhibitions to the public. Hunter was aware that his profession was not always appreciated by what he referred to as the "unenlightened" public. In his lectures he bemoaned living in a country where "liberty disposed the people to licentiousness and outrage, and Anatomists are not legally supplied with dead bodies." Due to this situation, he told his pupils that "particular care should be given to avoid giving offence to the populace." They must be on guard and speak with caution of "what may be passing in the School, especially with respect to dead bodies." It is necessary, Hunter concluded, to "shut our doors against strangers, or such people, as might chuse to visit us from an idle, or even malevolent curiosity."[9] In effect, however, except for some attempts to stop students from bringing their friends to the ever-popular lecture on the anatomy of the female genitalia, medical and nonmedical mingled together in Hunter's dissecting rooms at the Royal Academy of Art and in his spectacular anatomic museum. Until well into the nineteenth century, medical education was not well structured and there was no stable, set audience for medical texts or specimens; hence the division we make between the medical profession and the lay public would have struck Hunter as artificial and contrived.

Hunter does not mention a religious purpose or give a religious explanation for his work as an anatomist. But like Gunther von Hagens, Hunter attempted to instill awe in his audience at the wonders of the human form. He was a member of the Church of England. Although von Hagens is a self-avowed atheist, it is intriguing that he and Hunter use similar terms to describe the experience of dissecting and preparing specimens.[10] It is a holy event, a wondrous act for the future, and a glimpse into the creative universe.

While Hunter and von Hagens have provoked feelings of joy and revulsion in equal measure with their exhibits, they, like many anatomists before them, are credited with having few feelings. Hunter was described as a dispassionate workaholic: "He never married; he had no country house; he looks, in his portraits, a fastidious, fine gentleman; but he worked till he dropped and he lectured when he was dying."[11] Similarly, family, colleagues, and students of Von Hagens have described him as a detached workaholic.[12]

In fact, Hunter struggled with his emotions all his life. He was squeamish of operations. John Hunter revealed, in a biography of 1784, that his brother "at first . . . practiced both surgery and midwifery, but to the former of these he always had an aversion *because he hated operations, would often faint at an operation, even disliked to bleed, although he studied how the art might be improved.*[13] What was it Hunter "hated" about operations? The quotation suggests he had the fairly common phobia of turning faint when seeing blood flow from a body. Interestingly, Gunther von Hagens suffers from hemophilia and also fears the flow of blood.[14]

Without a doubt Hunter would admire the technical prowess and lifelike poses of *Bodies Revealed*. He sought to achieve this himself. Hunter purchased the body of a man before he was hanged for smuggling. The condemned man used Hunter's money to pay for his time in prison (rent was charged in the eighteenth century). Hunter received delivery of the robber's body from Tyburn Fields (a farm outside London where public hangings were held five days a week—it is now known as Oxford Street). The smuggler's body was still warm and, vastly excited by this good fortune, Hunter "was seized with the idea that the body might first be put in an attitude and allowed to stifen in it."[15] With the aid of the sculptor Agostino Carlini, Hunter quickly placed the body in the pose of the classical statue *The Dying Gaul*. He then had his students strip the flesh off the man and make a wax mold of the body.

The desire to acquire, and to display, and to keep experimenting with new techniques to preserve bodies is evident in *Bodies Revealed* and in Hunter's great work—a massive obstetrical atlas published in 1774. *The Anatomy of the Human Gravid Uterus* describes the course of pregnancy in words and in large, stunning engravings. Hunter had worked hard and paid a great deal to collect thirteen different dead women at various stages of pregnancy. He used the data he gathered from dissecting them to represent one normative pregnant woman. In a letter to a fellow Scottish anatomist, Hunter gloated over obtaining yet another specimen for his atlas:

> On the 11th of February I was so fortunate as to meet with a Gravid Uterus, to which, from that time all the hours have been dedicated which have been at my own disposal. I have been busy in injecting, dissecting, preserving, and shewing it, and in plan-ning and superintending drawings and plaister casts of it.[16]

Another link between *Bodies Revealed* and eighteenth-century anatomy is that stories abound of Hunter's willingness and desire to go beyond the boundaries of the medical and enter into what later became the world of the freak show. At the request of Martin Van Butchell, a former student of his who became a well-known dentist, Hunter embalmed Van Butchell's wife, Mary, and then Van Butchell kept her in a glass case in his drawing room. At set times each day, any member of the public could pay to see Mary. When Van Butchell remarried, his new wife, Elizabeth, insisted on having the drawing room to herself, and her husband reluctantly gave his first wife to the anatomy museum at the Royal College of Surgeons. Satirical verses circulated around London as to how Hunter had managed to produce such a lifelike Mrs. Van Butchell:

> To do his wife's dead Corps peculiar Honour
> Van Butchell wish'd to have it turned to stone,
> Hunter just cast his Gorgon looks upon her,
> And in a twinkling see the thing is done.[17]

Sadly, Mrs. Van Butchell's embalmed corpse was destroyed in the bombing of the Royal College of Surgeons during World War II.

Given his sensibilities, William Hunter would likely have been an enthusiastic supporter of public displays of plastinated bodies. Hunter would also have understood the controversies over such exhibits. In his day, as in ours, audience responses to such exhibitions varied. Some saw them as macabre and disgusting; others saw them as wondrous and awe-inspiring. The famous historian Edward Gibbon, for example, would not leave London during the winter of 1777 because he was attending William Hunter's lectures for two hours daily, "which have opened up to me a new and very entertaining scene within myself."[18]

CHAPTER EIGHT

Vive la differénce

Gunther von Hagens and His Maligned Copycats

LINDA SCHULTE-SASSE

I N THE PAST FEW YEARS, the myriad anatomic exhibitions sweeping the United States have come under increasing scrutiny for dubious ethical practices and a rampant commercialization of human remains preserved through plastination. Curiously, however, the enterprise that started it all in the first place, Gunther von Hagens's *Body Worlds*, seems all but exempt from this criticism. After being introduced to the United States through a wildly successful run at the California Science Center in 2001, *Body Worlds* (as well as its variations *Body Worlds* 2 and 3) has been hosted by the most prestigious of American museums, lauded by ethics review panels for its inspirational and educational value, and extolled by the American press; in short, it has become the most successful—and lucrative—exhibition ever. Not surprisingly, Von Hagens laments the proliferation of competitive anatomic shows as the product of "greed" in which a worthy venture has been "hijacked by corporate interests." *Body Worlds*' elaborate website explicitly labels rival shows "copycats" and charges them as inferior ethically (they display the bodies of executed Chinese prisoners rather than of consenting donors) and aesthetically (they "plagiarize" von Hagens's "unique expressive style" and offer "inferior imitations").

Few in the American press have found reason to question these assumptions. When, in 2008, ABC's *20/20* did a feature likening body exhibits to horror films, for example, it focused largely on *Body Worlds*' competitor Premier Exhibitions, giving von Hagens a forum to distinguish his own work as categorically different. It even showed him wiping away tears when recalling that he once received some bodies of executed prisoners, which he claimed to have cremated. In short, it would appear that, while *Bodies Revealed, Bodies, the Exhibition,* or *Our Body: The Universe Within* is to be shunned as morally reprehensible, commercial, and exploitative, *Body Worlds* is the exhibition of choice for the viewer seriously interested in pondering (according to its website) "difficult philosophical questions" about mortality.

It is ironic, then, that in von Hagens's native Germany and in Europe, where he first launched his shows, he has been plagued by attacks remarkably similar to the ones faced by his "copycats" today. If the *New York Times* decries Premier Exhibition's use of bodies of executed Chinese prisoners for commercial gain (November 18, 2005), the headline of German weekly *Der Spiegel*'s feature article of January 19, 2004, read "Dr Death: The Horrific Dealings of the Corpse-Exhibitor Gunther von Hagens." If *20/20* expressed shock that bodies are imported for Premier Exhibitions under the label "plastic models for teaching," the *Frankfurter Allgemeine Zeitung* sardonically reports that von Hagens's bodies have made their way into Germany as "animals unsuitable for human consumption" (January 8, 2000). Although Premier supplier Corcoran Laboratories is accused of the selling of human remains to private parties, *Spiegel* reports that von Hagens offered transparent horizontal slices of humans for 1400 to 2800 euros, depending on the durability, until 2008, when he ceased because of a public outcry. And if public officials like California Assemblywoman Fiona Ma seek to regulate bodies exhibitions in parts of the United States, virtually every city in Europe has tried to prohibit *Body Worlds.* Von Hagens founded his latest factory and theme park, the Plastinarium, in the economically depressed Guben, Germany, in 2006, only after the City Council of Sieniawa Zarska, Poland, refused him (and his former SS officer father, who was to direct the factory). *Spiegel* quotes a Polish official saying "the days when you could make soap and lampshades of human skin should be gone forever."

Seen from this point of view, von Hagens may be distinguishable from his competition for technical and entrepreneurial innovation, but not on qualitative grounds. Far from a violation of von Hagens's principles, the explosion of "copycat" shows attests to his success in following the logic of capitalism. Least of all would ethical purity set his enterprise apart. If there is one analogy consistently invoked by European critics, it is with the Holocaust. Andreas Nachama of the Berlin Jewish community labeled *Body Worlds* a logical consequence of twentieth-century history, Nobel laureate Günter Grass called von Hagens "Joseph Mengele 2" (referring to the infamous Auschwitz doctor), and the German horror film *Anatomie* (2000) fictionally linked von Hagens's plastinates with atrocities of the Third Reich, to cite a few examples.

But we take a different view in the United States and, even at this writing, *Body Worlds* is flourishing in new incarnations in Houston and Salt Lake City. How are we to understand this disparity in critical response to von Hagens in the United States and abroad, and how—given our growing scrutiny of the copycats—does *Body Worlds* stay so "clean," even if it launched what many perceive to be a disturbing trend—if it is, as the website boasts, the "original"?

Crucial to its affirmative reception in the United States has been *Body Worlds*' "step up" in venue from exhibition hall to museum. In Europe the exhibition, like the copycat shows running in the United States today, was housed in Berlin's Postbahnhof, Munich's Olympic Park Arena, or Vienna's fairgrounds—the sites of commerce-oriented exhibitions and rock concerts. In Hamburg in 2003 it appeared in an erotic art museum on the Reeperbahn, the city's red light district, accompanied by press slogans like "ultra-naked, uncensored"; in Brussels, it was staged in a converted slaughterhouse and meat market. In the United States, *Body Worlds* found its way into high-profile science museums with public and private funding sources, stringent exhibition criteria, and advisory boards comprising local theological, medical, and ethics representatives: Los Angeles' California Science Center, Cleveland's Great Lakes Science

Center, Chicago's Museum of Science and Industry, St. Paul's Science Museum of Minnesota, and others.

In short, "our" *Body Worlds* has turned up in places where we'd expect to find serious science, in institutions we'd expect to enlighten and better us. And, indeed, the main argument for hosting *Body Worlds*—and for distinguishing it from rival shows—is that it's about education and not commerce and that it respects humanity and human rights. Of central concern has been the provenance of the bodies. To preempt controversy, the museums have taken great pains to assure that von Hagens's specimens are from consenting donors; the exhibitions stress the "gift" of donation and display enlarged reproductions of donor consent forms. Several museums have even conducted their own investigations to assure that death certificates match donor forms for each body in the show (see the California Science Center's Ethics Review Report).

Yet the German press is far less convinced than our museums that von Hagens's body procurement has always been aboveboard. *Der Spiegel* reports that his plastination "factory" in Dalian, China (which used to be his main center of operation and is today the source for Premier Exhibitions' bodies), is near detention camps where many prisoners are executed and that he purchased bodies of mental patients from Novosibirsk. Von Hagens secured an injunction against *Spiegel* for these accusations but has constantly been dogged by similar charges and church-organized protests; as a Lexis Nexis scan of headlines in major German periodicals reveals, coverage of him nearly always involves investigations of ethical abuse and even the fraudulent use of the title Professor.

Whether or not all of the German allegations are true, one might question the scope of the American ethics investigations. As Lucia Tanassi points out, these have limited themselves to following the "paper trail" of each plastinate in its respective museum but have not investigated the full range of *Body World*'s specimens or the history of von Hagens's operations overall.[1] *Body Worlds* assures only that its full-body plastinates are from consenting donors and admits that it "accepts donations of bodies that have been provided by survivors," as well as "unclaimed bodies . . . from government agencies such as the Social Welfare Office" (*Body Worlds* catalog, 30). Hence Tanassi questions whether the "gifts" *Body Worlds* depends upon do not include nonconsenting "donors." Even if we can assume that—precisely as a result of his success—each full-body plastinate we see in a *Body Worlds* exhibition was willingly donated, this assurance begs the question whether the museums should be absolved of any responsibility for endorsing the much murkier origins of von Hagens's practice or whether von Hagens should be absolved of responsibility for having jump-started a trend that 20/20 labeled "ghoulish." Is von Hagens a "hijacked" missionary or the sorcerer's apprentice?

The museum venue has also been instrumental in a new staging of *Body Worlds* that renders it a different show from the one originally seen in Europe. The European shows were generally set in large exhibition halls uniformly lit with bright light. The full-body plastinates stood on metal stands surrounded by green foliage that gave one the feeling of walking through a crowded park. In the US museum venues, contrastive lighting effects lend the displays a more aesthetic, theatrical quality. Illuminated cases and sculptures glow in an otherwise dark space. The lighting is crucial to the effect of the exhibition, as it draws the eye to particular artifacts and dramatizes individual body sculptures as well as display cases, heightening their appearance as objects of beauty. The specimens are arranged so as to grow in dramatic intensity.

In addition, in the United States the plastinates are surrounded by imagery and written texts (missing from the European version) that also condition audience response as much as the theatrical lighting. Reproductions of anatomic art dating back to the Middle Ages cover the walls or are printed on banners. The presence of the anatomic drawings contextualizes the *Body Worlds* specimens in the tradition of anatomy and anatomic art (including "high" art) and thus endorses von Hagens's self-representation as the descendent of anatomic pioneers like da Vinci and Vesalius. Prominent among the wall reproductions is Rembrandt's 1632 painting, *The Anatomy Lesson of Dr. Tulp*, in which the doctor dissecting a corpse wears a hat that is the model for the fedora von Hagens has made his trademark. Von Hagens consistently promotes himself in the context of Rembrandt's painting. A *Body Worlds* publicity brochure shows his portrait against the background of the painting, which also served as the backdrop of a 2002 public autopsy he performed in London. The exhibition associates von Hagens with Rembrandt (and other painters like Caravaggio) in subtler ways as well, for its dramatic lighting pattern of dark spaces punctuated by illumination echoes the chiaroscuro technique of baroque and Renaissance masters. The exhibition space feels like a living painting.

The exhibitions in the United States also surround the exhibited bodies and body parts with language. Next to each full-body plastinate is a text panel explicating its anatomic features—an innovation of the advisory boards. But this "scientific" voice is not the only one speaking in *Body Worlds*. First, wall texts thanking the donors for making our experience possible give the exhibition an "authorial" voice, a "we" that expresses thanks to "them." Second, quotations by famous thinkers and philosophers allow an array of canonical voices to reflect on the nature of humanity, counteracting the pure physicality of the cadavers and anchoring *Body Worlds* in a tradition of thought that mirrors the iconic tradition invoked through the anatomic drawings. Psalm 8 ("Thou made [man] little less than God, crowning with glory and honor over all thy creatures") reassures the visitor that, even if it breaks a taboo, the exhibition is consistent with Judeo-Christian ethics. Von Hagens's own words stand next to Goethe's, implicitly granting him the stature of the great thinkers represented. Finally, isolated nouns describing humanity— courage, fortitude, faith, compassion, spirit, justice, and so forth— loom high in the exhibition space, their vertical layout almost suggesting an advertising slogan (as one wonders what happened to greed, gluttony, sadism, and, indeed, voyeurism).

One might wonder why a scientific exhibition needs such an abundance of "voices." Only the anatomic explanations tell us what to look for: the thank you notes, philosophers' quotes, and nouns listing human virtues are irrelevant other than to tell us *how* to look—with reverence, and not with voyeuristic pleasure. In other words, the changes in *Body Worlds*' staging suggest the need to construct a context that conditions viewers and forestalls a potentially "inappropriate" response, stifling qualms with an omnipresence of "taste." While in Europe the exhibition hall venue and more obvious consumer orientation rendered *Body Worlds* an "event," a freak show, here it is cloaked in a mantle of solemnity, respectability, and high culture, assuring that *Body Worlds* is a celebration of enlightenment, scientific progress, and altruism. (It is telling that, while the US *Body Worlds* is a favorite site for school field trips, the province of Brandenburg, Germany, prohibits such trips.)

In short, a central reason why Americans have not found *Body Worlds* more troubling, despite a strong religious presence in our culture, is the discrepancy between its practice and its discourse. The display of cadavers strips away our illusions about ourselves and shows us as pure matter. Yet the show's visual and verbal "language" frames the demystified body in a humanism that compensates, that makes us comfortable. Before we can object—as have so many Europeans—that the display of filleted corpses violates human dignity, the show answers: "No, we are all about your dignity! How can we not be with all these great philosophers and a picture of Rembrandt on the wall?"

Moreover, *Body Worlds* successfully mobilizes aesthetics in the service of science, and not vice versa. We have seen how the exhibition's newfound respectability depends on its aesthetically pleasant "artsy" presentation—yet this respectability depends on art being subordinated to science, to the "educational reason" for the art, as the California Science Center's Ethics Report put it. Since von Hagens not only styles human cadavers as sculptures but signs them, it is hard to imagine this practice escaping controversy but for its claim to scientific and educational value. Given the widespread attacks on art that challenges taboos (Robert Mapplethorpe's photography,

Chris Ofili's *Holy Virgin Mary*, Andres Serrano's *Piss Christ*), can one imagine a display of real human art-corpses in an American art museum meeting with such a friendly reception?

Our validation of displayed cadavers in the name of science, combined with the limitation of controversy to the issue of donor consent (however important), may cause us to overlook the cultural implications of *Body Worlds* and, indeed, all of the shows that imitate its artistic style. Several critics have already noted the abundance of "conservative" social messages in these exhibitions. *Body Worlds'* authoritative "voices" in the wall quotations are all Western and male, the number of male plastinates far outweighs the number of women, and when women are presented they tend to assume classic female poses: the ballerina, the skater held high in the air by a virile male, and the archer. Many female plastinates, like the reclining pregnant woman, highlight the woman's reproductive organs. As Megan Stern points out, the exhibition adheres to the "longstanding assumption within anatomical tradition that the male represents the anatomical norm and that the female is of interest primarily as a means of demonstrating the reproductive system."[2] While the reclining pregnant woman has been one of the few controversial specimens in *Body Worlds*, the controversy has surrounded the display of a dead fetus and largely ignored her sexualized body stance, one redolent of the tradition of the nude in European painting. Her pose harks back to classical art, recalling Giorgione's or Velázquez's *Venus* or Ingres's *Grand Odalisque*.

Perhaps most striking is *Body Worlds'* idealization of bodily perfection. A preponderance of the plastinates are athletes, and sequencing tends to be important as we progress from the Smoker, whose charred lungs show us the perils of bad habits, to the glory of the Blocking Goalkeeper, the Archer, the Balance-Beam Gymnast, the Hurdler, and the Figure Skating Pair—idealistic models of what we'd like to be but probably never will. The rationale for this presentation is an encouragement of wellness and "lifestyle choices." But the celebration of man perfected has a less flattering legacy: this kind of aesthetic was the major tenet of Nazi art, which privileged sculpture above other art forms. As Klaus Wolpert stresses in his study of Nazi art, the monomaniacal representation of the naked body was one of the few distinguishing traits of National Socialist sculpture; it celebrated the body "untouched by the stigma of decay or 'degeneration,' of the ugliness of everyday life."[3] Von Hagens's human statues have transparent affinities to the sculptures of premiere Nazi sculptor Arno Breker, as well as to filmmaker Leni Riefenstahl's athletic figures in *Olympia* (1937). Alternatively, von Hagens's ideal reproduces the imagery of consumer capitalism with its focus on a bodily perfection achieved by estrangement from the self. While von Hagens reconfigures his bodies as art, postmodern subjects nip, tuck, and Botox themselves; in both cases the body is alienated, aestheticized, and commodified—in short, incrementally dehumanized. In an essay aptly titled "Shiny Happy People," Megan Stern comments how *Body Worlds'* durable and hyper-real specimens promise "what the real body can never deliver." Imogen O'Rorke likens von Hagens's "necro-bodies" to Barbie dolls, which likewise present an unachievable and durable ideal in plastic form, while Carl Elliott likens von Hagens's sculptures to taxidermy.[4]

In addition to worrying about *whether* these "shiny happy people" have consented to their postmortal plastination, one might also scrutinize *why* they do so (today von Hagens has a successful donor program). The museums may extol the donor's "generous gift" at the cost of "personal tragedies," but on the multiple choice form where donors check reasons for their decision, only one choice echoes the assumption that donors wanted to leave their bodies to the ubiquitous "good cause." All the other reasons listed on the donor form are self-oriented: some are pragmatic (no funeral costs), and others suggest the desire to live on in some form, even a kind of euphoria in being on display (word choices include "excited" and "fascinated"). *Die Zeit* cites interviews with donors who report a desire for a life after death that is happier than life itself (August 21, 2003), and *Body Worlds'* website convenes future donors to discuss their "post mortal lives." As Megan Stern puts it, "plastination offers a uniquely secular, material form of immortality . . . *Body Worlds* is, quite literally, a consumer heaven."

The exhibition's discourse projects onto the donors a messianic mission that may have little to do with their "gift." It is one thing to give one's body or organs so that someone else may live and quite another thing to allow one's body to be appropriated as a work of art produced and signed by a living artist, to be reincarnated as Gunther von Hagens's Hurdler in his commercial blockbuster *Body Worlds*. And while the "respectful" treatment of a dead body is culturally arbitrary and there is nothing new in the display and aesthetic refiguring of bodies, there is something new in this particular form of commercial and artistic appropriation of human remains. The fact that the donors agreed to be part of this, "wanted" it, should encourage discussion, not stifle it. People also "want" to estrange their own bodies by appearing on *Extreme Makeover* or *The Swan*; they "want" to appear on *Big Brother* or be humiliated on *American Idol*. In other words, "consent" and commodification are by no means mutually exclusive.

If, as *Body Worlds* assures us, the donors' identities and "tragic destinies" remain respectfully sheltered from our ogling gaze, this is certainly not the case for the auteur von Hagens, whose eccentricity is becoming legend and whose ego is enhanced as he signs post-human artworks, poses with them, or rides in Berlin's Love Parade disguised as them. A hagiography on von Hagens by his wife and "friends," *Pushing the Limits*, stylizes him as a pioneer and freedom fighter, highlighting the narrative of his imprisonment after an attempt to flee his native East Germany. This stylization extends to the link the *Body Worlds* website makes between von Hagens and James Bond, occasioned by the appearance of plastinates in the 2006 Bond film *Casino Royale*: von Hagens and others behind the Iron Curtain were "inspired" by Bond's "anti-authoritarian and unconventional" style in his "archetypal battle between good and evil." (See chapter 9 for more on James Bond.)

I hope to have shown how the contrasts between the reception of

Body Worlds in Europe and in the United States, as well as between *Body Worlds* and its copycats, bring to the fore issues that a "tasteful" presentation clouds. The more *Body Worlds* highlights its uniqueness vis-à-vis the "other" shows, the more obvious is its sameness as part of a cultural phenomenon and a consumer enterprise. If, as it claims, *Body Worlds'* mission is the democratization of science, what's the point of even talking about "copycats"? Isn't it a good thing if more people experience the "gift" of plastination, as long as the rules are followed? Isn't it macabre to claim sculpted human bodies as somebody's "original," as if they were Michelangelo's *David*? Perhaps von Hagens's copycats do us a favor in making more transparent the central question raised by these exhibitions, irrespective of venue or tasteful presentation: When we start using cadavers as objects of art, commerce, and spectacle, do we run the risk of forgetting that somebody can easily become some body?

CHAPTER NINE

Normative Objections to Posing Plastinated Bodies

An Ethics of Bodily Repose

TARRIS ROSELL, DMIN, PHD

U PON LEARNING THAT the *Bodies Revealed* exhibit is coming to town, colleagues gather around a conference table at the Center for Practical Bioethics and a spirited dialogue erupts. One of our staff members has been recruited to serve on an ethics advisory group for the science museum sponsor. We discuss the various ethical controversies that surround this exhibit. Our moral hackles are raised by allegations of ethical impropriety in the procurement of bodies. Some of us are put off by the sensationalist billboard photos of plastinated corpses in athletic poses. A few are bothered by the utilization of an Asian ethnic minority group for the educational or entertainment needs of Westerners.

Several around the conference table say that they will not visit the exhibit and do not think the ethics center should be involved with it in any way. Others think that we should organize educational activities to discuss the moral controversies.

During exhibition months, I present *Bodies Revealed* as a case study for classes of seminarians studying Christian ethics and medical students studying bioethics. The latter are young adults primarily from Christian faith communities. All of the theological students are Christian and most tend toward middle age, with equal numbers of women and men. Some are liberal and others are conservative on matters of morals. In both classes, discussion is lively regarding the morality or immorality of exhibits of plastinated bodies.

Only a few persons in either class have visited this exhibit or intend to do so. One medical student who had trained and volunteered as a docent expresses distress at what she experienced during approximately forty hours on site. She has a hard time articulating what exactly seemed "unethical" about *Bodies Revealed* but recalls an incident when an adolescent boy was caught taking cell phone photos of a female cadaver's genitalia. The student docent had sensed what she terms a patent disrespect for the dead by some, not all, visitors. Initially, other students are unable to specify further what seems morally objectionable, yet something about this exhibit just doesn't feel right to most of these very bright scholars. Is it merely what someone describes as the "ick factor"?

Class discussions produce a minority opinion that is voiced in support of *Bodies Revealed* on grounds of educational value, freedom of expression, and marketplace principles. Anyway, who is harmed? The dead themselves? Postmortem rights or harms seem unlikely to this group. Even the dissenting majority can think of potential alterations to the current exhibition that might make it "okay," or seem less wrong, anyway. They suggest the involvement of nonprofit entities only, fully documented and informed consent by the donors of the bodies, diversity of ethnicity in consented corpses, clearly educational intent and not that of entertainment, avoidance of sensationalism in advertising, and federal regulation of international trafficking of bodies or body parts for display or sale.

A Christian clergy colleague, my former student, also engages me in conversation about *Bodies Revealed*. Jack has seen the advertisements but doesn't plan to go visit.[1] He is a particularly thoughtful young man, one with a congenitally gentle and calm demeanor. I have never before heard Jack raise his voice or seen him get visibly angry. But when I ask him what he thinks about this interesting exhibit and ethical controversy down at the Science Museum in Union Station, my friend responds immediately and irately, emoting more than thinking. Jack's voice shakes a bit as he states rather loudly, "Let me tell you this much: If it was my wife's body on display down there, I'd burn the place down!"

After weeks of moral waffling and collegial dialogue, I decided to go see the exhibit. I had hesitated to go mostly out of respect for the dead. In the end, I went out of respect for the dying. My eighty-one-year-old mother had just been diagnosed with stage IV colorectal cancer with metastases to the liver, lungs, ovaries, and peritoneum. After receiving this sobering news in a phone call from home, I felt a need to see those body parts in context, not only in photographs online, but in a human body if possible. The exhibit of plastinated bodies made this possible. I bought one of the last tickets, during the last week of several months of opportunity, and arrived only an hour or so before closing. A bioethicist without benefit of a bio-medical education, I wanted desperately now to try to understand something of what was happening to my mother's body, to gain a layperson's appreciation of what would be happening to her in months to come.

I went, looked, pondered, learned, and wondered. As every anatomist discovers, the human body is breathtakingly beautiful, complexly exquisite. The plastination process itself is ingenious, worthy of the awe its products engender. The *Bodies Revealed* exhibit seemed to me then less ethically problematic than I had imagined. The bodies on display did not have the disgusting sensationalism of some billboard advertisements for the exhibit. Exhibitors clearly had educational intent. Everywhere there were descriptions, explanations, definitions, revelations—the anatomy lessons for which I had come. And there were other lessons more explicitly moral, especially that of a highly promoted display of tar-black lungs. A bin of cigarette packs was immediately adjacent and nearly filled by impulsive disposals by visiting smokers under sudden conviction of sin.

Much of what I experienced there was surely good. Yet moral ambivalence remains, in part because of some deeply embedded practices of my own faith tradition in regard to how we treat the dead body. Norms of respect underlie the objections I note to current practices of plastination exhibits, but the focus here is slightly different than that of others who write in that vein. I am interested in normative disposition of human remains and especially in Western Christian concepts of terminal or temporary "rest."

Plastinated Bodies and What We Normally Do

We do interesting things with the bodies of our deceased, from burying to burning to boxing in aboveground mausoleum crypts. Some people donate their bodies to science or to medical education. Others donate body parts for transplantation, enabling hundreds of thousands of us to live longer with better quality of life.

In most cases thereafter, human remains (even *cremains*, as cremated bodies are called) are disposed of in some ritually prescribed way. Whatever remains of a person after death is ultimately "laid to rest." That is how we say it in moral communities of my upbringing and residence. In the funerary rituals I have attended or officiated at over several decades throughout the United States, the language we use describes a state of "rest," conveying what I will call here a normative state of "bodily repose."

Plastinated bodies on exhibition tours are treated differently. This is a new thing in ways potentially more significant than mere technique or technology, artistry or entrepreneurism. What strikes us upon gazing at these decedents on display is that many have been preserved as bodies in frozen motion: running, biking, jumping, batting, dealing, shooting, kicking, throwing.[2] They are the playing or working dead, perpetually posed without repose, positioned without disposition. For those who subscribe to an ethics of bodily repose, this treatment of human corpses is morally repugnant, a violation of unwritten but communally understood notions of right and wrong.

An Ethics of Bodily Repose

Western Christian traditions are varied in regard to most matters of life and death and doctrines of life after death. But there is a striking consensus on respectful treatment of the dead body, inclusive of disposition in keeping with something like rest. Funeral and memorial services incorporate readings from the Bible, some of which foster the belief that death is a state like that of sleep, perhaps temporary until some eschatological event involving bodily resurrection. There are dozens of biblical references to death as sleep or rest. Christian scriptures commonly quoted include the following: "Blessed are the dead which die in the Lord from henceforth: Yea, saith the Spirit, that they may rest from their labors, and their works do follow them" (Revelation 14:13).[3] "And all wept and bewailed her: but he [Jesus] said, Weep not; she is not dead, but sleepeth" (Luke 8:52). "Now is Christ risen from the dead, and become the first fruits of them that slept" (1 Corinthians 15:20).

Hebrew Bible (Old Testament) readings at Christian services of remembrance and burial supplement those of the New Testament and reinforce notions of restful disposition: "And many of them that sleep in the dust of the earth shall awake, some to everlasting life, and some to shame and everlasting contempt" (Daniel 12:2). So the Psalmist prays: "Consider and hear me, O Lord my God: lighten mine eyes, lest I sleep the sleep of death" (Psalm 13:3).

Prayers and other ritual language of the traditional Christian funeral impress upon mourners that their loved one's "work is now ended," "his labors are over," "she has run the race and finished the tasks set before her." The "dead in Christ" are finally "at rest." The liturgy for services of mourning includes phrases like these: "Jesus Christ, who is the resurrection and the life . . . also hath taught us . . . not to be sorry . . . for those who sleep in him; . . . that when we shall depart this life, we may rest in him."[4] "We have in remembrance before thee all who lived this life with us, . . . and who are now at rest with thee . . . We bless thy name for all thy servants who have kept the faith, and finished their course, and are at rest with thee."[5]

As mourners file past an open casket at a "viewing" or "visitation" or "wake," the informal gathering that precedes the funeral, one overhears utterances that "she looks so peaceful" and that, "if you didn't know better, you'd think he was just sleeping." Funeral services, and those of Christian worship generally, often include a hymn that begins, "For all the saints who from their labors rest."[6]

Teen-aged mourners of a classmate or relative adapt an ancient tradition when writing on the windshield of their vehicle the name of the deceased, followed by "R.I.P.": "Rest in peace." Out at the cemetery, this same inscription will be found on weathered tombstones. Of medieval Christian origin, the Latin phrase "Requiescat in pace" is really a brief prayer on behalf of the decedent: "May he rest in peace."

At the graveside ceremony, a clergy officiant speaks words of committal, "Ashes to ashes, dust to dust," before the casket or urn is lowered or placed and the parishioner is finally "laid to rest." Whatever hardships might have taken place during the decedent's lifetime, there comes a time at the end to "lay it all down." If interment had been delayed slightly due to organ and tissue recovery or for a lengthier period after whole-body donation for medical education and research, there is a sense during the committal service that the donor receives at long last a well-deserved quietus. Perhaps especially if a body had gone missing for a time due to war or murder, survivors breathe a sigh of relief at their loved one's return to a "final resting place."[7] Even a beloved pet euthanized for the relief of suffering is said euphemistically to have been "put to sleep."

The Christian dead are said both to be "asleep in Jesus" (1 Thessalonians 4:14) and yet also "present with the Lord" in heaven (2 Corinthians 5:8), referencing New Testament passages that are somewhat ambiguous as to the immediate state of spiritual existence after physical death. Doctrinal beliefs on these matters vary from church to church and even among members of the same congregation. Among some Christian sects going back to early centuries of the Common Era, there has been a doctrine referred to as "soulsleep."[8] The belief is that, when people die, they enter into a sort of spiritual unconsciousness until being roused by God to awareness once again so as to enter heaven or hell. Scriptures that are used to support this teaching include Daniel 12:2, quoted above. The account of Jesus's crucifixion in the Gospel of Matthew contains a sentence that is also referenced by those who understand death as soul-sleep: "And the graves were opened; and many bodies of the saints which slept arose, and came out of the graves after his resurrection" (Matthew 27:52–53). Regardless of soul-sleep beliefs, the ultimate hope for many Christians is that death's spiritual rest will eventuate in bodily resurrection, being raised up "to sit together in heavenly places in Christ" (Ephesians 2:6).

Christologies, speculative or dogmatic teachings about the nature of Christ, embrace various interpretations of resurrection. All are grounded in a New Testament Greek term, *anastasis*, which literally means "a rising up" or, in its verb form, "to stand up again." Stories of Jesus Christ tell of one who dies, is "laid in a tomb," and then, after three days of interment, emerges alive from the grave (Matthew 28; Luke 24; John 20). This event is referred to as Christ's *anastasis*, resurrection, because the act of going from the sleep of death to the resumption of corporeal life necessarily involves a shift of position from that to which the dead normatively are disposed. The resurrected Jesus Christ "stands up" from his dispositional state of bodily repose. Followers of Christ then might hope someday to do the same: "But I would not have you to be ignorant, brethren, concerning them which are asleep, that ye sorrow not, even as others which have no hope. For if we believe that Jesus died and rose again, even so them also which sleep in Jesus will God bring with him" (1 Thessalonians 4:13–14; see also John 11:25; 1 Corinthians 15; 1 Peter 1).

If the dead are believed to be at rest or even "asleep," then it is understandable that believers would inter the body in a manner that coheres with belief. The living normally rest lying down; therefore, the dead too are normatively "laid to rest." To do otherwise would simply be unthinkable. If someone nonetheless were to do the unthinkable, adherents to an ethics of bodily repose would wonder why. Violation of the norm surely appears morally suspect and garners a normative objection: "This is wrong."

Counterexamples to the tradition of laying a body illustrate the strength and centrality of that tradition. Violating it is something that is done to enemies rather than friends. When the Hebrew King Saul dies in battle, Philistine captors come to "strip the slain": "And they cut off his head, and stripped off his armour, . . . And they fastened his body to the wall of Beth-shan" (1 Samuel 31:8–10). This disdainful treatment of the corpse produces distress or outrage within a moral community committed to an ethics of bodily repose. There is an immediate response to rectify the wrong: "And when the inhabitants of Jabesh-gilead heard of that which the Philistines had done to Saul, all the valiant men arose, and went all night, and took the body of Saul and the bodies of his sons from the wall of Beth-shan, and came to Jabesh, and burnt them there. And they took their bones, and buried them under a tree at Jabesh" (1 Samuel 31:11–13).

Centuries later, Jewish Christians wrote of a righteous rich man who risked life and reputation so as respectfully to inter the body of Jesus in a state of repose. Following crucifixion on a tree, the corpse remained scandalously and publicly displayed in its upright position.

> And after this Joseph of Arimathaea, being a disciple of Jesus, but secretly for fear of the Jews, besought Pilate that he might take away the body of Jesus; and Pilate gave him leave. He came therefore, and took the body of Jesus. And there came also Nicodemus . . . Then took they the body of Jesus, and wound it in linen clothes with the spices, as the manner of the Jews is to bury. Now in the place where he was crucified there was a garden; and in the garden a new sepulchre, wherein was never man yet laid. There laid they Jesus (John 19:38–42; cf. Mark 15:42–47)

Objections to Objections

It is one thing to describe what is normatively done and quite another to assert that it ought to be so. I have not made that claim, preferring to describe and not prescribe an ethics of bodily repose. If prone to moralistic prescription, one might argue an "ought" for some particular community and still not make a universal claim. Hence, even if it were agreed that Western Christians seem obligated to dispose their dead in a resting posture, that is no mandate for other moral communities. It is possible even to maintain that respectful disposition of human remains is a prima facie duty (and to ground this in something other than the contested notion of postmortem rights), yet not make universal claims regarding what constitutes "respectful" treatment of the dead. Indeed, there are tribal peoples of the Amazon for whom familial cannibalism is the definitive sign of respect.[9] Mummification practices of some ancient cultures occasionally depicted decedents as active not sleeping, working instead of resting.[10] Cremation, less contentiously, raises another potential exception to Christian rules of repose.

Decedent cremation rather than decomposition has been practiced by many moral communities for centuries, of course, and has gained popularity during the past century in both Europe and North America. For example, cremation rates in Arizona increased from 60 percent to 68 percent in one year,[11] and nationally they are now greater than 26 percent (compared to more than 45% in Canada), according to the Cremation Association of North America.[12] Cremains ultimately might be disposed through burial, scattering, crypting, placement on a mantel, or occasionally in ways more creative, as in urns that double as a lamp or clock.

Interred ashes probably do not carry the same sense of repose that is gained by an intact corpse horizontally disposed. Still, the liturgical language of our mourning rituals is similar or identical for cremations and burials. In either case, the decedent is "laid to rest." Hence, cremation as practiced by Christians in the West is not entirely an aberration from the dispositional ethics I delineate. So, too, we read that the non-normative cremations of the desecrated bodies of King Saul and his sons culminated with normative disposition of repose via burial of their bones "under a tree" (1 Samuel 31:13).

It could be conjectured that Western Christians typically inter in a state of bodily repose merely because this seems actually to be the physiological disposition of the deceased. They literally have ceased their labors, no longer getting up to go work or play. Yet there must also be a perceived value in disposing the body accordingly; otherwise mourners would not pay considerable sums to have morticians prepare the body in a manner suggesting sleep and then spend thousands more to view their loved one in that posture prior to the final "laying to rest" in burial. This sort of preparation of the body is done sometimes even before cremation. Given this communal value and these prevalent customs, new practices— that is, of noninterment and action posing of plastinated bodies for traveling shows— stand in stark contrast to typical Western cultural and Christian norms. Exhibits like *Bodies Revealed* may appear, then, to be ethically wrong.

The response of my friend Jack to the thought of his wife's remains being plastinated and displayed suggests the depth of emotion that such treatment of the dead might evoke. "I'd burn the place down!" Since Jack is a person who is not much prone to anger and not at all prone to violence, his response—even as a thought experiment—is noteworthy. It indicates a deep moral disturbance and an understandable expression of outrage in reaction to the imagined disrespect of the body of a beloved spouse signified by plastination. This outburst may be interpreted as evidence of an embedded ethics of bodily repose. It seems better to Jack that his wife's plastinated body be incinerated in an act of arson than publicly and perpetually to be posed in some immoral and uninterred display. His hypothetical act of criminal cremation would depose the body to a normative state more fitting to human decedents of our moral community. His beloved would be laid to rest.

I presume a case can be made that everyone, everywhere and always, ought to do whatever dispositional actions are deemed culturally respectful of the bodies of persons deceased. Principles and practices derivative of the universal remain particular, however, and applicable only to the moral communities from which they arise. Such norms are in flux, more or less. They could change as a result of communal experience, empirical evidence, or persuasion.

It may be that exhibits of plastinated bodies will induce that kind of change in contemporary Western societies. But public consternation expressed to date suggests that such a culture shift has not yet occurred here, or not entirely so. Rather, there is evidence that a sense of moral repugnance remains widespread despite large turnouts and profits generated by the several exhibitions on tour. One way of understanding these normative objections to the posing of plastinated bodies is via explication of Western Christian norms of decedent disposition, that is, an ethics of bodily repose.

CHAPTER TEN

For Ronnie and Donnie

MYRA CHRISTOPHER

W HEN I WAS GROWING UP in the late 1950s and early 1960s,
there was a special day each October. On that particular day
children across the city were dismissed from school and given a free
pass to attend the State Fair of Texas—the largest state fair in the
country. My friends and I, unaccompanied by adults, went to Fair
Park in Dallas, Texas, to enjoy the fair.

It was usually the same week as the "Red River Shootout," when
the University of Texas and the University of Oklahoma played
football in the Cotton Bowl, but our presence had nothing to do
with the game. We were supposed to go to the fair so that we could
take advantage of the educational opportunities. There were won-
derfully educational exhibits inside beautiful art deco halls that had
originally been built for the Texas centennial in 1885. But the real
attraction for me and my friends was the midway.

The midway was our idea of heaven. There were dozens of great
rides. The ones I remember most vividly are the giant wooden roller
coaster, the tallest Ferris wheel in North America at 212.5 feet, the
scrambler (where my best friend, Patricia Davis, broke her collar
bone when we were in the eighth grade), and the "Super Himalaya,"
where songs by Danny and the Juniors played endlessly. There were
also games of chance to be played and huge stuffed animals to be
won for those who had lots of money to venture. If you could knock
down the right number of bowling pins with a baseball, toss three
rings over a peg, or pick up the rubber duck with the right number

on the bottom as dozens floated by, you could have your choice of a stuffed animal that was almost as big as you were.

And the food—god, it was fabulous! Everything you ever wanted and your mother didn't want you to have was right there: red hot dogs, greasy hamburgers, clouds of pink cotton candy on a paper cone, fried corn on the cob, blocks of ice cream on a stick dipped in chocolate and rolled in crushed peanuts, and the coldest Cokes and Dr Peppers imaginable. My favorite lunch was a Fletcher's Corny Dog accompanied by a paper cup full of real French fries with lots of ketchup and a big glass of ice-cold lemonade.

It wasn't all heavenly. The midway had a dark side. Past the roller coaster, there were freak shows. There were tents where you could see a two-headed cow or a live headless chicken. Those were just gross. More troublesome to me was a show where you could meet Ronnie and Donnie (conjoined twins) and a huge tent that housed the Circus of Strange People and Freaks. The barker outside the Circus told the crowd that, for the price of admission, you could see a legless man walk on his hands and do pull-ups, the tattooed lady with paintings *all over her body*, a real giant—more than seven feet tall, the lobster boy with claws for hands, and the rubber woman—a contortionist who would *amaze and excite* you.

Every year my friends would goad me into buying a ticket, and I would stand in line with them until it was time to climb the wooden stairs and enter the tent. Then I would chicken out. I just couldn't do it. I never went inside. It just seemed wrong.

I hadn't thought of those shows for more than forty years. But I remembered them when the exhibit of plastinated bodies, *Bodies Revealed*, came to Kansas City last year. I was aware of these exhibits of plastinated bodies. I had been in other cities when they had been present. I had read about them and heard people talk about them. Now I was being asked by friends, family, and colleagues if I planned to attend. I was clear about my answer: I would not. I was not as clear about a response to their next question: "Why not?"

Initially, I thought I might get off the hook by referring to the wisdom of repugnance, or, as Leon Kass referred to it, the "yuck factor." But I wasn't sure that my colleagues would allow me to claim any sort of wisdom based on my response to the billboards that were plastered all around town. I attended a seminar about the ethical issues raised by the exhibit and listened carefully to the discussion that followed, when several concerns were raised. The most frequent objection raised by those present was basically the sanctity of the human body (i.e., that our bodies are given to us by God and are not ours to do with as we please).

I am not religious. I describe myself as a positive agnostic. So I couldn't base my objections on arguments about the sanctity of the human body. I worried about the informed consent of the donors, but even if the documentation of voluntary consent were impeccable (which it was not), I knew that I still would not have attended. I was concerned because the bodies in this particular exhibition were all identifiably Asian. But if the bodies had not been ethnically identifiable or had been multi-ethnic, I still would not have gone to Union Station to see the exhibit. So I began thinking about what lay beneath my clear and strong resolve—to try to discover within me something more than an emotional or aesthetic repulsion—to develop a rational argument.

I am a strong proponent of autonomy. I support the right of women to choose abortion and that of competent adults to refuse life-sustaining medical treatment. But I am not a libertarian. I believe there are limits to autonomy and to our right to do as we wish with our bodies. I absolutely abhor tattoos and have always feared that one of our daughters might decide to get one; still, I would defend her right to do so. I want women to have access to birth control pills, and I believe *Roe v. Wade* is a reasonable legal construct for the permissibility of abortions. But I'm opposed to prostitution—legal or otherwise. I want medical students to have access to cadavers, but I am delighted that most medical schools now have somber ceremonies acknowledging the significant contribution made by the individual who donated his or her body when the cadaver is no longer useful. I have participated in the decision for a deceased loved one to become an organ donor and regret that so many people die on transplant waiting lists, but I oppose free markets to create an incentive for organ donation. So, from a secular perspective, how do I sort this all out?

If we think of the body as property—and if we believe that the principle of autonomy gives us the right to do what we please with our bodies so long as the act does not harm another—then why not let people choose plastination rather than burial or cremation? I think the answer lies somewhere between Socrates and Kant.

When Socrates was charged with corrupting the youth of Athens, he argued that he couldn't possibly be guilty because that would lead to him living in a corrupt society and would be contrary to his own interests. If we don't treat the human body as something more than physical property, we act against our own self-interest and create a situation that undermines civil society. Imagine a society in which we treated the human body as property (even with protections for children and the cognitively incapacitated) and, other than the consent and contract requirements of law, removed all restrictions on the buying, selling, or altering of bodies and body parts. I don't think anyone would refer to it as civil or just. It would not be a place I would want to live.

However, as I have thought about my discomfort with the *Bodies Revealed* exhibit and others like it, my fundamental objection is that, in my opinion, they diminish human dignity by objectifying the people whose bodies are displayed. They show disrespect for the persons whom the "donors" once were. I do not believe that our requirement to respect the dignity of the individual stops when the individual dies. If we didn't respect the dignity of the dead, we would not have the elaborate rituals that we have for burial or cremation.

Recently, the television program *Entertainment Tonight* aired photographs of the dead rock star Michael Jackson as he was loaded into an ambulance. *OK Magazine* published such a photo on its cover. People were outraged. Various websites were abuzz with complaints about the lack of respect and privacy and, yes, even the loss of dignity for the King of Pop. We should be just as outraged by the undignified public display of an anonymous person from China.

CHAPTER ELEVEN

The Creeping Illusionizing of Identity from Neurobiology to Newgenics

BARBARA MARIA STAFFORD, PHD

No, it had to be sweet
as grass, the kind of stuff that's habit-
forming like all things half-conceived:
for instance, Adam
anesthetized, and God, part surgeon, part
cosmic dating service,
taking her out for the first time
to see how it would go (the Bible leaves this part out,
although the Greeks not believing
in premature withdrawal, left it in)!

ELEANOR WILNER, "CANDIED"

*B*ody Worlds: The Anatomical Exhibition of Real Human Bodies—Gunther von Hagens's hugely popular show of flying, jumping, chess-playing plastinates—has captured our attention at a peculiar moment in both the history of art and the progress of science. These "real human bodies" are rendered dry and odorless by a preservation process that replaces fluids with silicon rubber. Wet bodies and slimy organs look leathery and flexible. Because of their uniformly desiccated surfaces and the absence of both slipperiness and the stench of corruption, the treated bodies appear uncannily individual and ideal at the same time.

Exhibiting human remains (ranked PG-13) has been a controversial and much debated practice, especially in the bioethical and anthropological communities.[1] These voided human specimens offer an occasion to reflect not only on the oddly depersonalized and distributed conception of the contemporary body but also on the irony that the real slides into the unreal when it is hygienically processed as if it were a computer-assisted vision. I propose the paradox that these hyper-real anatomies are unimaginable without the backdrop of new electronic media and the concept of both science and art as disembodied information that such imaging technologies foster.

Significantly, and not unlike elaborate biomolecular procedures, the elaborate process of dismembering and then re-engineering formerly intact human specimens—so that they appear vertically "exploded" or horizontally expanded (as in running) or enacting a role, such as a leaping ballet dancer or a dribbling basketball player—takes place primarily in remote facilities, in this case in China and Kyrgyzstan. Equally remote is the description of the complex preservation technique. For example, we are informed with clinical detachment, and without accompanying illustration, that the decomposition of the corpse is stopped with formaldehyde or freezing. It is "then either dissected or sawed into slices, depending on how it will be permanently preserved. Frozen body fluids are replaced by acetone in a frigid (minus 13 degrees) acetone bath. Most specimens, particularly bones and intestines, must be defatted in room temperature acetone before plastination can begin. In a vacuum chamber, the acetone is squeezed out of the specimen and gradually replaced by plastic," with each whole body requiring "up to 1500 hours to prepare." This passage is maddeningly vague, though precise.[2]

The explicit violence associated with the venerable tradition of the Western public autopsy—underlined in Vesalius' disemboweled cadaver stretched out on the dissecting table in the frontispiece to his *De humani corporis fabrica* (1543)[3] or sublimated in Frederik Ruysch's late seventeenth-century allegorical tableaux of aborted fetuses dressed in the lace of their afterbirth—is still implicit here, but buried. As is true of biocomputing, the plastinates can theoretically be made of any material as long as certain principles are fulfilled. It just so happens that these particular artificial agents are compiled from human flesh. Such abstraction serves to detach the bodies we are looking at from the lengthy tearing down and building up material operations they underwent to achieve their final appearance. No wonder, then, that, despite their reality as technically accurate images, they do not possess the three-dimensionality, that is, the physical actuality and cognitive awareness, of seventeenth- and eighteenth-century wax models from Zumbo to Fontana. No "I" or self-consciousness ever inhabited the trepanned cranium. The motion of the limbs still occurs automatically, even without the intervention of the will. Unlike the reanimation and re-formation promised at the Last Judgment, this collaging and montaging of stripped down bodies happens in silicon.[4]

Von Hagens, then, is very much of the tradition of Western art and very much of the moment in his deconstruction and reconstruction of organisms that are organic-inorganic chimeras. But he has nudged paradigms in both science and art, creating a strange and disorienting sense of creep. Creep, scientists tell us, is "the slow deformation of a material that occurs when it is held under a constant load, such as the gradual stretching of a piano or violin string."[5] During creep in metals or metallic alloys, voids form and grow with time, accompanied by fundamental changes in the original crystalline structure. This essay is about a different kind of resistless transition—not in brass but in biomedicine and brain science—whereby the organism is being "deformed" by computational science and "knowledge engineering" into a synthetic medium. I suggest a parallel—what this steady veering by the life sciences toward the model of information-processing is doing to our everyday attitudes about the body, from genetics to consciousness.

Since Darwin, creep has been a crucial concept in biology. It has become increasingly difficult to uphold any concept of species that insists on the permanence and stability of each kind. Contemporary biology seems smitten by visions of transmutation and transmutability far exceeding those proposed by Lamarck or Geoffroy de Saint Hilaire. Modern biology first underwent the "biochemical" turn and then the "molecular" turn; now biology is in the process of making the "compositionist" (synthetic) turn—raising deep classificatory, epistemological, and even ontological questions not just about the reducibility of the living to chemistry or physics but about whether the phenomenon of life itself can be engineered.[6] It is not unusual to hear about "custom-building biological systems,"[7] and some have ambitious plans "to modify the whole behavior of the cell" by engineering it from the ground up "rather than tinkering with a handful of genes or tweaking a metabolic pathway or two, as do today's genetic engineers."[8]

During the past four decades, cloning has become merely the most visible of an encroaching series of human genetic engineering techniques. These sophisticated biological tweakings range from somatic engineering (or the intentional transformation of genes in the body) to germline engineering (extending to the descendants of a person) through chemical manipulation. Already animal lovers, with the aid of a paw-sized piece of tissue and $32,000, can replicate their deceased cat using chromatin transfer, which has a much higher success rate than the nuclear transfer method used to clone Dolly.

In this reckless company, Craig Venter's neo-Darwinian aspiration to assemble a Whole Earth Gene Catalog by collecting the DNA of everything on the planet does not seem extreme. He imagines a "combinatorial genomics" using existing robotic technologies not only to collect 100,000 new species and tens of millions of new genes but to construct a functioning synthetic genome and to dream up new life forms.[9] William Hurlbut's proposal to grow teratomas (naturally occurring monsters that grow from an egg and sperm cell but without the balance of gene expression to create an integrated organism) as a way around the stem-cell stalemate sounds bizarre but creepily plausible.[10] Such a horrible hodgepodge of organs might never constitute a true embryo destined for a human trajectory, but the fact that scientists would be engineering mutant nonembryos to satisfy theologians ought to produce more than repugnance.

While human cloning may not yet be technically feasible, fear continues to be expressed in many quarters about the possibility of genetically engineering people and their offspring to make them bigger or stronger, age more slowly, be less aggressive, or exhibit more intelligence. We are now entering an age of self-eugenics in which improvements to human bodies and minds are becoming a primary goal in biomedical research and clinical treatment. A major contributing factor to this state of affairs is that DNA sequences are treated less as molecules and more as information. Between 1863 and 1868, the heredity theories of Herbert Spencer, Gregor Mendel, Charles Darwin, and Francis Galton—when taken together—helped bring about a convergence of social planning, philosophy, and biology.[11] Such evolutionary concepts culminated in an ideology to improve the human race by the careful breeding of the "best" people. They did not result in the revolutionary notion, now common, of constructing oneself into a wholly new and superlative specimen through tools, techniques, and products available on the private market. This privatized self-retooling is dominated by capital investment, patent applications, and trade secrets. Genetic elitism—linking evolution, heredity, and scientific and technical progress to the most intimate aspects of life and intelligence—has given us "newgenics." As Lori Andrews notes, recent legal disputes about what kind of substance human genes are devolve on whether they are identified primarily as raw material or as patentable subject matter, like software.[12]

By one view, genes merely constitute "a library of defined components that can be assembled into control systems for biological computation, or used to program bacteria in order to produce interesting proteins and other compounds."[13] Timothy Lenoir was prescient in his claim that, in the ivory tower, "experiments in silico" are rapidly overtaking experiments in vitro.[14] But his prediction that there may be no laboratory to which molecular biologists may return has taken an unexpected twist.

Evidently, the era of postacademic "garage biology" is now upon us, touted in the popular press as an "art" that can be learned. This entrepreneurial enterprise can apparently be taught to robots or sold as a toy to children ten years and older ("The Discovery DNA Explorer Kit"). Biotechnology is pushing universities and industry closer together commercially. The intellectual world in which knowledge was gained by interrogating the world is being replaced by one that only interrogates an encroaching, encompassing, and world-flattening Web.[15]

The ongoing efforts at reconstructing our morphological and genetic identity are not just the stuff of academic laboratories and research. They encroach upon the clinical arena, where interventions ranging from elective cosmetic surgery, to growth hormone injections, to the dream of the perfect child, to the specter of re-designing humans, all intermingle.[16]

In this peculiar social climate where everything and nothing seems individually possible, it is not surprising that consciousness, too, has become something "distributed," dispersed, depersonalized. The mind is seen to be not just an internal information-processing device (neural network) but a circuit that ineluctably spreads out into the environment to which it has been correlated and for which it has been made. Just as the individual self is taken to be a kind of intellectual property coded in gene sequences and disseminated in gene pools, self-awareness is distributed across spaces, times, participants, and objects.[17] Like the use of the passive voice in writing, this way of imagining our most intimate self conceals agency, work, and responsibility and turns the individual into an object that has things done to it.[18]

Paradoxically, at the same time that somatic self-production is on the rise, health care is being outsourced, handing the parcelized and powerless patient over to automated expert systems. If everything from the identification of an illness to diagnostic consultation is considered to be disembodied "information," then "medical management" seems an apt term for the industry concerned with analyzing, compiling, and re-engineering the data of any ailing organism.[19] The nineteenth-century proponents of measurement, statistical accuracy, and standardization would surely have marveled at our unjaundiced trust in blanket quantification.[20]

My intention is not to mount a critique of the textuality of scientific nomenclature. Instead, I want to contribute to the imaging side of current studies on the metamorphosis of biology and, indeed, many other disciplines, into an information science. Evelyn Fox Keller, Donna Haraway, and Richard Doyle have examined both the cultural and the economic context of using linguistic metaphors such as "DNA code," the "genome library," "text," "translation." More recently, Timothy Lenoir compellingly addressed the theme of "tools to theory," that is, instrumentation, software, and the role of the computational medium itself in the historical (from the mid-1960s onward) theorizing of biology as an information science.[21] Similarly, in the area of design theory, John Thackara claims that we have been relentlessly filling the world with connecting devices—from body implants to broadband communications to smart appliances—without first discussing the purposes they might serve.[22]

How is this expanding scientific process of abstraction and fragmentation changing our conception of human identity, cognition, emotion, and behavior? Do DNA identity, genetic identity—like credit card identity or ethnic identity—also exert pressure on the social sphere? I am thinking of identity politics, where the demand for personal recognition is based, not on shared human attributes, but on respect for the individual as different.[23] Stating it otherwise, if the demand for a politics of recognition in the age of globalization is not about achieving eventual inclusion within the fold of humankind, how will proximity-making electronic media, systems, and services manage to rescue the marginal from being rolled into one common, universal narrative? Conversely, how will genetic testing avoid categorizing us by our unique and deep propensities? What will happen when we physically and mentally become what we create?

The *Body Worlds* exhibitions speak directly to these anxieties in ways that are best illustrated, perhaps, by their phenomenal popularity. The exhibition has already attracted more than 30 million viewers in Europe and Asia since the 1995 premier in Tokyo. Some twenty-five freestanding, whole bodies, ranging from obese adults to toned athletes, are skinned, dissected, laid bare. These rotless cadavers demonstrate everything from clogged arteries, blackened lungs, arborescent capillaries, and tangled nerves to prosthetic knees. While these flayed, living statues are not supported on pedestals or enveloped by protective vitrines—making it easy to inspect them close up and in the round—an array of sliced and sectioned livers, kidneys, gallbladders, spleens, hearts, and brains, as luminous as stained glass—are deployed in backlit cases. A separate, curtained-off area contains a panorama of developing embryos and fetuses, which enframe a reclining woman. Her exposed womb contains an eight-month-old baby, snugly nestled amid the intestines.

Both the surrounding wall text and the press packet vacillate between telling a cautionary tale of self-abuse (through smoking or bad diet) or, conversely, potential self-improvement and a more uplifting general educational message of providing insights into human anatomy. Taken together with the cleaned-up visual evidence, these combined discourses add up to a powerful rhetoric of self-fashioning. As I suggested, this re-engineering of the natural body into a synthetic more-than-body cannot be separated from the biomedia of self-eugenics. It is not accidental that von Hagens has compared the reconstitution of the limited and flawed self of the donor, as well as his supersized re-scaling (as in the bicyclist expanded to one and a half times normal size), or even creation de novo (the fictitious Mystical Plastinate), to the perfected identity bestowed on the patient by the plastic surgeon.[24] But, surely, growth hormones and nanotechnology might also be invoked.

Was Donald Winnicott correct? Does reality gain intensity from surviving continuous destruction? The famous object-relations psychologist was speaking of the difficulties attending the recognition of an "other"—seen as external and beyond one's personal omnipotence (and thus subjected to continual unconscious destruction). Significantly, this stranger is brought into our consciousness only if capable of contributing something new to our lives, that is, enduring only to the extent that he or she becomes of "use."[25] I want to use the importance given to the role of form in Winnicott's theory about prolonged birth memory—the relief that comes "when the end is in sight from the beginning"—as a springboard to think about Gunther von Hagens's real specimens. Why could these purified units, existing outside the distortions of the mind, come into existence only against the backdrop of medical and research genetics?

Just as cloning, protein sequencing, and gene product amplification are dependent on synthesizing vast quantities of molecular data into symbolic textual representations, magneto-encephalography and electromyographic scans visually record the massive flow of neuronal information within the nervous system. Unlike the illustrated anatomies of Vesalius or Albinus, however, biocomputers have gotten us deep inside the skinless human animal with the explosion of data on DNA, RNA, and protein sequences. But what we, in fact, see are simulations: biofictional ciphers generated and analyzed through automated tools utilizing intensive computer calculations and elaborate search algorithms.

By contrast, in the early modern medical world, the human body was considered variable and mutable. Up to the eighteenth century, its humoral properties contributed to the belief that it was porous and always in flux. As a microcosm of the surrounding universe, it continuously interacted with that greater, God-given macrocosm to which it more or less perfectly corresponded, depending on fluctuating conditions of health or sickness. What gave the human body its specific open and fragile identity, then, was the ongoing need to establish an analogical bond linking its particular psychosomatic components with those of external nature—whose matter was also shot through with numinousness.[26] Thus, from Vesalius to Vicq d'Azyr, cadavers were never lifeless but were transformed (in myriad illustrated anatomy books) into energetically striding, standing, or sitting figures. Even when recumbant, they were never merely prone but were actively being eviscerated or mined for bones, muscles, organs, and tissues that demonstrated the glory and nothingness of human material on an amphitheatrical stage, under the watchful eye of the public.

That lengthy era in which you could learn about biology only by studying natural systems has finally drawn to a close. Eugene Thacker has argued that each of the new biotechnologies I mentioned earlier in this essay articulates a specific kind of body. These "biomedia," further, are predicated on the premise that computation is biology, that there is a fundamental equivalency between the biological and digital domains so that they can be rendered interchangeable in terms of materials and function.[27] It is this rapidly becoming self-evident fact of the "technical reconditioning of the biological" that, I believe, we clearly see in Hagens's direct reworkings of gross anatomy. The throngs attending this display of "Anatomy Art" can, if they so wish, experience in the exhibition, and in common, what is difficult or impossible for ordinary persons to see in the biomolecular body, or the adult stem cell body, or the nuclear transfer cloned body. That is, the average, interested citizen can witness and think about the consequences of instumentalizing biology into designed contexts.

Craft and Narrative in *Body Worlds*

An Aesthetic Consideration

NEIL A. WARD, MFA

D R. VON HAGENS has not asked us to evaluate him as an artist. In fact, he has not asked us to evaluate him on any terms. Like Dr. Kevorkian, he seems offended by any inference that he may be courting our, or anyone's, approval. Instead, he poses himself as a reality in the landscape that we must deal with, not pass judgment upon. He seems rather like a refugee from an Eastern Bloc state who cannot understand why democratic institutions dedicated to personal freedom are so determined to hamstring themselves—Ayn Rand, reborn and redux.

He has outrun most of the allegations of his having neglected the legal details of body donors' "informed consent." Still, the idea of establishing a registry of donors—who would approach a pregnant women with an aneurysm about plastinating her and her fetus's bodies in the event of complications?—defines insouciance beyond the common limits, even while it allows him to sneer at the hamfisted protocols of competing plastinators, the subpoenas of various countries' attorneys general, and the hand-wringing of some bioethicists.

This atmosphere of taunt seems to be daring us to evaluate him. The *Body Worlds* exhibit and other displays of plastinated bodies invite evaluation in aesthetic terms, beyond a more narrowly defined legal or moral framework. Aesthetics is ethics—concerned with decisions based on value judgments. An aesthetic critique is a crucial element of the bioethical response to von Hagens and his work.

Body Worlds: Is it Duchamp's Urinal?

In what sense should we understand *Body Worlds* as a work of art? For most people, including me, it is conceptual art. But conceptual art itself is the subject of definitional ambiguities. "In conceptual art," writes Sol LeWitt, "the idea or concept is the most important aspect of the work. When an artist uses a conceptual form of art, it means that all of the planning and decisions are made beforehand and the execution is a perfunctory affair. The idea becomes a machine that makes the art."[1]

Conceptual art, in LeWitt's sense, cuts off the idea of "art" from its roots in "craft," or technique. The usual assumption is that mastery of a craft liberates deeper visions or insights than could emerge lacking the rigor of such a mastery. Francisco Goya (for example) might neither have been able to record, *nor to experience*, the visions of his *Desastres de la Guerra* etchings of the Peninsular War but for his rigorous mastery and application of the aquatint process; the hand guides the eye. Conceptual art says, Not necessarily. The artist—the most *authentic* artist—bypasses the mere hand, generally shouting "look, Ma!" while doing so. The presumption of conceptual art is that the work will insult or outrage folks, or possibly just bewilder them, in a novel or interesting way and touch off some broader concatenation.

Fountain, by Marcel Duchamp (1917), defines the genre. This work consisted of a mass-produced urinal, which the artist—known then as a pioneering cubist—purchased from its manufacturer, signed "R. Mutt," inverted, and mounted at an exhibition in New York. I say "consisted" because *Fountain* is no longer extant; the gallery seems to have thrown it away afterward. Some doubt whether it was actually ever publicly viewed. Nonetheless, it was photographed, and several artists created facsimiles in porcelain (one is now in London's Tate Modern museum), and the critical buzz has persisted to the present. A poll of "500 leading art critics" cited by the BBC in 2004 named *Fountain* the "most influential" piece of modern art ever, just ahead of Picasso's *Demoiselles d'Avignon.*

The goal of creating and displaying such a piece is to release a narrative. This idea has not gotten old as fast as one might have supposed. Generations of artists since Duchamp have refined and expanded the genre. Von Hagens clearly belongs in this group, even though his work does not have the perfunctory execution that Le-Witt identified as a stipulation. In that, von Hagens is not alone. Stravinsky labored at *Le Sacre du Printemps*, Beckett at *En Attendant Godot*, and Buñuel at his *L'Âge d'Or*, but all these works nonetheless caused the kind of bewilderment or riots characterizing true conceptual art. Similar claims might be made for the photography of Diane Arbus, Robert Mapplethorpe, and Andres Serrano. Today, not only the "idea or concept" of these works has obtained validation but also the considerable craft that went into their execution. In a similar way, the early anxiety about photography's lack of human involvement and this gap's troublesome implications, especially for portraiture, has been largely dispelled.

Body Worlds: Is it Warhol's Soup Cans?

Dr. von Hagens can justifiably claim to have mastered a craft; so could Andy Warhol by the time he grew restless as a commercial artist and decided to display beautifully rendered images of soup cans and soft drink bottles as fine art and become the fundamental element in a realignment of taste in the avant-garde. (That same poll of five hundred critics calls Warhol's *Marilyn Diptych* the third "most influential" work of modern art.)

The analogy between Von Hagens and Warhol is extensive: both show a clarity of execution, both attract reproaches for banality and commercialism, and both wield some influence as rebellious pundits from a perch attained through practical success. Warhol has come down to us as the exponent of shallowness as a function of mass production, and his "fifteen minutes of fame" coinage has taken such hold that it has, perhaps, already disassociated itself from its late creator for this generation.

But what is von Hagens dishing up, and from what basis in craft does it derive? It is clearly not simply from the craft of plastinating bodies. Von Hagens and his wife, Dr. Angelina Whalley, who is billed as "creative and conceptual designer of *Body Worlds*," have exercised aesthetic choices beyond, and specifically alienated from, the craft implicit in Dr. von Hagens's rigorous development of the plastination technology. In this, they stand in contrast to other innovative artist-craftsmen such as Goya, whose mastery of a craft, in itself, appears integrated into his vision.

Although von Hagens's apprenticeship in plastination extended over decades and was material and rigorous, the exhibits as we have them seem, at best, tangentially related to that particular journey. Rather, we sense in the exhibits a taste and mode of selection with deep autobiographical referents—the experience of the war refugee, the hospitalized hemophiliac, the alienated student, the ransomed political prisoner, and finally the embrace of the West, and that embrace's intriguing reciprocity.

For the *Body Worlds* publicity team—apparently Dr. Whalley is in charge—the choices are pedagogic ones: *Body Worlds* displays figures in the poses of living people engaged in recreational activities—playing poker, swimming, dancing, horseback riding—in order, says the website, "to illustrate different anatomical features, and to help the viewer relate the specimen to their own body." Familiarization—fair enough. Seeing a plastinated body stretch or lift, or appear to express will, evokes a sympathetic response and helps the audience bring the experience of a plastinated muscle into its own experience.

But more subtle than familiarization is demystification, which is to knock the body off its pedestal as cradle of the soul. This challenge is packed with "narrative," in the tradition of conceptual art, in that it poses an open question to most traditions in Western religion and at least some in Eastern: if the soul is immortal and the body is transient, what is the relationship between the two? Which forms the other? Is the Platonic assumption of body-soul dualism merely a facile proposition?

Von Hagens seems to be challenging our ability to push aside the question, to compel us to admit that any concept of "soul" we possess is, in the end, rooted in materiality; for we cannot conceptualize "soul" as truly dematerialized. Or, at least, neither Milton nor Dante could, and no one else has made nearly as game an attempt. In our own time, the late physicist Richard Feynman has outlined the difficulty of forming analogies to properties entirely alienated from material experience. As Feynman explains, trying to visualize the "electric and magnetic field" as the workings of invisible angels is merely begging the question, "No, it is not like imagining invisible angels . . . Why? Because invisible angels are understandable."[2] As long as we can only visualize "almost-invisible angels," that limitation itself sets the limit of our ability to analogize.

Religious leaders—not yet the pope—have taken up von Hagens's challenge, although with no undue rigor and little enough traction, positing again that the body is the cradle of the soul and is, for that reason, due a certain reverence, which the *Body Worlds* exhibit insensitively violates. The Right Reverend Nigel McCulloch, bishop of Manchester, has protested to the Manchester Museum of Science and Industry "the effect public anatomy shows have on public opinion" and fears that von Hagens's works may depress donations to transplant banks.[3] Nor have civil executives been silent. President Hugo Chávez of Venezuela (responding not to the *Body Worlds* exhibit but to one of the competing ones) is even less nuanced: "We are in the midst of something macabre. They are human bodies. Human bodies! This is a really clear sign of the huge moral decomposition that is hitting our planet."[4] Carcasses unwelcome in Caracas; permit withdrawn. Thus far, the conceptual artist is running rings around the authorities.

Body Worlds: Danse Macabre

Since President Chávez brings it up, we must acknowledge that the late medieval danse macabre motif, with its lively, worldly, irreverently grinning skeletal figures, clearly inhabits the *Body Worlds* vision. The religious or spiritual meaning of this motif (German *Totentanz* and English "dance of death") has attracted little agreement among either religious or art historians. Some have posited that these compositions—which began appearing shortly after the Black Death of the mid-fourteenth century had run its course in Europe— were a means of comforting, or providing a homeopathic tonic to, a population that felt itself vulnerable to sudden and arbitrary death. Others suggest that the motif reinforced the separation of the body and soul by imposing on the skeletons so many worldly attributes, particularly dancing and laughing, which would supposedly not convey along with the soul after death.

The danse macabre's most lineal survival to our own age is Mexico's *Día de los muertos*, an idiosyncratic celebration of All Saints Day, associating remembrance of the dead with skulls and skeletal objects, icons of a material sense of death. These objects—jewelry, candle holders, and so forth—are ubiquitous in the society throughout the year, not just in the vicinity of the holiday. This usage seems to conform to our former sense—comforting a population by inoculating it from the ubiquity of arbitrary death. But the Protestant Reformation's fascination with the iconography (both Dürer and Holbein produced engravings, for popular circulation) seems to associate the danse macabre with the separation of body and soul. Unwilling to assimilate sexuality through southern Catholicism's pre-Lenten carnival, the early Protestants reinforced the notion of mortifying the flesh by disassociating bodies and their functions with the soul, drawing an association among the flesh of fornication, the flesh on the butcher's block, and the flesh of the putrifying corpse.

Not much sets *Body Worlds*' use of the danse macabre icon apart from either its Catholic or Reformation invocation—"irreverent" body poses included—except for the exhibits' generally secular context. If the exhibits were mounted in churches, they would attract one inference. If they were mounted in Las Vegas gaming halls or strip clubs, they would attract another. But the preferred venue is the respectable, even somewhat philistine, municipal museum of science or natural history, supported by graphic posters on buses and at subway stations. In the tradition of conceptual art, this approach manages to comment upon and critique religion and science at the same time. Which, it asks, is which, today, or does it matter?

Von Hagens's corpus of work, in any case, is not "putrifying"; it has fulfilled the dream of the pharaohs. Disputing the notion of Ambrose Bierce, who termed life "a spiritual pickle preserving the body from decay,"[5] von Hagens's corpses neither decay nor smell and perhaps invite speculation from those raised in any Western religious tradition as to whether the spirit or soul these corporeal figures have "exhaled" can claim to have survived in eternity as well as the body did.

Whatever the East German anatomist's religious education or whatever his idiosyncratic life's choices actually respond to, it is nearly irresistible to speculate that the terms of his youthful experience in the post-Fascist miasma of the socialist state in some way freeze-dried his sensibilities and replaced their organic component with a polymer of semi-processed Kant and Nietzsche; that he is less the "Dr. Frankenstein" some of the tabloids have dubbed him than he is the creature itself. Those who have (as I have) puzzled over Antonin Artaud's demand (1938) that artists be "like victims burnt at the stake, signaling through the flames" may find it to the point to consider von Hagens's craft and aesthetic sense in light of that demand.[6]

P. T. Barnum Weighs In

I live in Washington, DC, the location of the Walter Reed Hospital, which for several generations has hosted an educational museum, now known as the National Museum of Health and Medicine. It is located a good distance from the National Mall, so its treasures are somewhat less known to the touring public than to the resident cognoscenti, some of whom speak of having had a moment of private enraptured enlightenment inside the place. One may view the cancerous lung and the diseased liver (indeed, von Hagens also displays entries in these categories), but what seem to send viewers into deep, introspective meditations are the pickled elephantiasis leg and the giant hairball (was it really extracted from the stomach of a girl who compulsively chewed on the ends of her pigtails?).

But the Walter Reed exhibit, and similar exhibits in other cities, such as Philadelphia's Mutter Museum and Chicago's Museum of Science and Industry, cannot shake a cultic, old-fashioned, even dowdy reputation. They are decidedly out of your face, not in it. Von Hagens has been determined to step beyond the presumptions and taboos that attend these decades-old predecessors.

The tension between the public's need to connect with the basic realities of mortality and the options that have arisen from other sources to process those realities—religious, social, patriotic, or what have you—resolves in the form of taboos. Body taboos govern diet, sexual behavior, excretory behavior, and the treatment of illness, among other matters. In our society, one common alibi for evading these taboos is "education." The most obvious example is the policy of *National Geographic* magazine, at least until recently, to allow the titillation of children in middle-class households by publishing photographs of full nudity, as long as a "cover" of anthropological education was provided.

The carnival freak show, although no longer quite the commonplace it was a hundred or two hundred years ago, commonly evaded a disreputable odor by posing as educational, with the hermaphrodite, dwarf, quadruple amputee, or tattooed savage being exhibited by a barker who styled himself a "professor." P. T. Barnum, who did so much to quench the national appetite for this kind of close encounter, termed his touring variety show "Barnum's Grand Scientific and Musical Theater" and called his New York exhibit hall a museum.

We should view von Hagens's and Whalley's claims about the educational value of *Body Worlds* within this context. Educational value the exhibits certainly have, but Dr. Whalley's claims that "I try to present the body in a dramatic, memorable, beautiful way so that people can learn about anatomy, disease and health" and that "I have been able to educate far more people about health than I ever would if I'd been a surgeon" appear disingenuous, as does Dr. von Hagens's hope that the exhibits will attract more young people into careers in anatomy.

Rather, considered as conceptual art, *Body Worlds*' educational value lies in the larger purview of "narrative"—setting up dialogue loops about our internal assumptions about the separation of flesh and spirit, our prejudices about the moral vision of Eastern Bloc refugees or Germans' reputation for ideological absolutism, the basis of our notions of sexuality and religion, and our questions about the motives and choices of the terminally ill when they contemplate eternity. These matters—indelicate as they are—are by no means unworthy of meditation, and the indecorous forum von Hagens and his team have established to display them to us seems unlikely to exhaust its generative power in the artists' lifetime.

Afterword

Plastination's Share of Mind

NEIL A. WARD, MFA, *and* JOHN D. LANTOS, MD

I T WOULD BE UNJUST to the breadth of the positions and insights of our multidisciplinary contributors to close this book with an attempt to synthesize or summarize the observations presented. These twelve essays suggest that the issues raised by the plastination of bodies reach more deeply into the center of human concerns than one might at first expect. They also implicitly address a question that often arises in discussions of plastination: "Who cares?" Why, from among the many issues that claim the attention of the discipline of bioethics, should we concern ourselves with plastination? Does the plastination of bodies, as an issue bidding for a thinking human's share of mind, really bid quite at the level of such matters as stem cell policy, organ donation protocols, prenatal gender selection, prioritization of access to medical care, euthanasia, or—that hoary old grandma of the culture wars—abortion? The list of bioethical issues that demonstrably affect the lives, quality of lives, and decision environments of millions could certainly grow. Indeed, it might be tempting to place the issues addressed in this book among the items of least concern. There is something decidedly low-brow about plastination. The odor of the carnival adheres to it. But that very observation invites a further reflection.

One thing that sets the plastinated bodies exhibits apart from other bioethics controversies is their popularity and notoriety. Not many manifestations of bioethics issues draw crowds in Las Vegas. That, in itself, might count as a point against serious attention to the plastination phenomenon. But another feature also sets the plastinated bodies controversy apart. For such a popular phenomenon, it is also surprisingly nuanced.

Indeed, other bioethics issues are "popular." Stem cells, abortion, and euthanasia, for example, are often in the headlines and are discussed on prime-time television. But those issues have a different sort of popularity. Discussions of those issues evoke strong views among laymen. The passions that surround the abortion debate, for example, have determined, to a large extent, the composition of the U.S. Congress and federal judiciary, to say nothing of state jurisdictions. Abortion as an issue, not necessarily as a procedure, is quite popular. But it is not nuanced. There is little more to be said, little that creative thought can add, to the crystallized positions held with more or less passion throughout our political and, indeed, ethical consensus.

In contrast, public and scholarly engagement with the issues raised by plastinated bodies remains vivid and fluid. The plastination of bodies has seized the popular attention not because it is cutting-edge science, nor because philosophers and theologians respond to it, but because the work itself is mentally engaging, technically impressive, ethically problematic, educationally presumptuous, and commercially successful. The issues raised by plastination open a window onto the larger set of issues surrounding any alteration, display, or trade in body parts. These issues grip and engage a share of the popular mind.

Views and values surrounding this late intrusion into our attention have imposed themselves at a point of fine balance. When we contemplate this controversy, we do not easily sink to an ossified formulation but continue to turn the matter over, attend to the perspectives of others, and ponder and work to unravel the actual stakes within contexts that have a certain refreshing unfamiliarity—as the essays in this volume, we think, demonstrate.

The notion of "popularity" itself tends to bring the plastination of bodies and their exhibition into a certain disrepute. Some scholars may raise their noses and question whether a book like this one merely gives legitimacy to a matter that belongs in the popular gutter. Let us acknowledge this common academic prejudice—and set it aside. The *Body Worlds* exhibits invite the layman to study, not just anatomy, *but the bioethics universe and his or her place in it*. Such popularizations raise questions and offer opportunities to democratize debates that are too often held in the rarefied air of academic seminar rooms or in the esoteric prose of scholarly journals.

Such venues do not do justice to the task. As medical technology and medical policy impose themselves increasingly within the fabric of our lives, everybody needs to obtain a certain literacy in the philosophic issues surrounding these phenomena. The public display of plastinated bodies offers a rich opportunity for public discussion that goes beyond the anatomy primer or the public health homily which the exhibits' brochures take pains to highlight. Perhaps, as those brochures claim, the exhibits can teach the public about the consequences of smoking and drinking, and if the plastinators follow the trajectory of anatomic museums throughout history, they will soon move on to explicit displays about the ravages of venereal disease.

But we are thinking of a different set of lessons: the venue for a democratized debate in the bioethics realm. Consider the well-publicized Terri Schiavo case, culminating in 2005, which touched off a lively debate within church groups, at workplaces, over Internet social media, and in many other popular fora about the ethics of assigning to a husband the decision of whether to pull the plug on his young wife who was (or was not?) in a persistent vegetative state. An act of Congress led to an accelerated Supreme Court review of the question—and a unanimous ruling, even within that ideologically divided Court, upholding that husband's right.

In bioethics circles, the Schiavo case was uncontroversial, and the Supreme Court ruling acknowledged the common law that had grown out of the bioethical consensus. But it *was indeed* controversial for a large cohort of passionately engaged laymen. The ethicists had, it seems, outpaced the body politic in "resolving" an issue outside of a frame of reference familiar and acceptable to the public.

The great health care reform debate of 2009–10 likewise pointed toward a parade of unexamined bioethical propositions for the attentions of battling factions. Death panels? Rationing? Health as entitlement? Had we the tools to frame up these questions, let alone debate them?

The existence of a rift between bioethicists and the public should concern us. Geoffrey Rees's essay (chapter 3) suggests that the ethical tradition of the Protestant Reformation implies the right of everyone to study anatomy for him- or herself. This view of the relationship between experts and the general public is the foundation of our political system. We have a right to know. But a right to know implies an obligation at some level to prepare the inquisitive mind for new and complex facts, new and complex moral formulations, new needs for resolutions, decisions, and responses to issues. Cultivating a bioethical awareness, through the medium of the controversy over the plastination of bodies, reaches beyond an abstract suite of philosophical notions and becomes an occasion for an inquiry into the history of moral thought, the shifting positions of churches, artists, and regulatory authorities, into the different ways in which commercial trade in bodies and body parts has been carried out in the past.

We need a laboratory in which everyone can investigate a bioethics agenda. The plastination exhibits *and the controversy surrounding them* are, in Rees's sense, something like a Book of Common Prayer—a text, accessible to all, which may be variously interpreted and which may offer glimpses of intrinsically unfamiliar states and choices.

We offer this probing and speculative volume as a commentary on the Book of Common Bioethics that the plastination controversy has set forth. We hope that this commentary will be of interest to those who go to a plastination exhibit and leave with a feeling of perplexity, to those who find themselves disputing the propriety of the pedagogy or the implicit theology of such exhibits, and to those who simply want to learn more about the history of anatomic display and the commercial use of bodies. As medical options and medical policy continue to enlarge their place in our society's core, engagement with the questions posed by *Body Worlds* may facilitate popular access to bioethical debate.

NOTES

Introduction

1. Schultze S, "Corpses May Revive Museum; 'Body World' Exhibit Expected to Draw up to 500,000," *JSO On Line*, www.freerepublic.com/focus/f-chat/1954028/posts. Accessed Nov 26, 2009.

2. American Association of Museums, www.aam-us.org/aboutmuseums/abc.cfm#visitors. Accessed Nov 29, 2009.

3. "How to Donate Your Body to 'Body Worlds' for Plastination," www.ehow.com/how_5490537_donate-body-body-worlds-plastination.html. Accessed Nov 27, 2009.

4. "Munich: Body Worlds Exhibit Unconstitutional," www.dw-world.de/dw/article/0,,764682,00.html. Accessed Nov 27, 2009.

5. Schulte-Sasse L, "Advise and Consent: On the Americanization of Body Worlds," *BioSocieties* 2006;1:369–384, http://journals.cambridge.org/action/displayAbstract;jsessionid=0912BC42228FA6998CE20D0021018EEE.tomcat1?fromPage=online&aid=570528.

6. Kolhekar A, "A Gross Indignity to Humans," *Science* 2003;5653:2070.

7. Wenig G, "Under the Skin: Is 'Body Worlds' Anti Jewish Values?" *Jewish Journal.com*, 12 August 2004, www.jewishjournal.com. Accessed Feb 26, 2011.

8. Archdiocese of Vancouver, "Concerns about Body Worlds," www.rcav.org/media/bodyworlds. Accessed Feb 26, 2011.

9. Phelan M, "Death, the Body, and *Body Worlds 3*," *Catholic Sun*, 2007, www.catholicsun.org. Accessed Feb 26, 2011.

10. Patterson T, "Body Worlds Impresario 'Used Corpses of Executed Prisoners for Exhibition,'" *Telegraph.co.uk*, January 25, 2004, www.telegraph.co.uk/news/worldnews/europe/germany/1452542/Body-Worlds-impresario-used-corpses-of-executed-prisoners-for-exhibition.html. Accessed Nov 27, 2009.

11. Ross B, Schwartz R, Schechter A, "Exclusive: Secret Trade in Chinese

Bodies," http://abcnews.go.com/Blotter/story?id=4291334. Accessed Nov 27, 2009.

12. "NY Reaches Deal with 'Bodies' Exhibitor on Corpses," www.the freelibrary.com/NY+reaches+deal+with+%60Bodies%27+exhibitor+on +corpses-a01611555838.

13. Kuhnel W, "Statement by the Anatomische Gesellschaft on the Infamous Body-World Show of Dr. Gunther von Hagens (2004)," www.ifaa.net/ PLEXUS_DECEMBER_2004-FINAL.pdf. Accessed Jan 2009.

14. Jury L, "Anatomy of a Row: The Bodies at an Exhibition That May Deter Donors," *The Independent*, 16 March 2002, www.independent.co. uk/news/uk/home-news/anatomy-of-a-row-the-bodies-at-an-exhibition -that-may-deter-donors-654178.html.

15. Jones DG, Whitaker MI, "Engaging with Plastination and the Body Worlds Phenomenon: A Cultural and Intellectual Challenge for Anatomists," *Clin Anat* 2009;22:770–776.

16. Park K, "The Criminal and the Saintly Body: Autopsy and Dissection in Renaissance Italy," *Renaissance Quarterly* 1994;47(1):1–33, www .thefreelibrary.com/The+criminal+and+the+saintly+body:+autopsy+and +dissection+in+...-a015488510. Accessed Nov 27, 2009.

17. Hansen JV, "Resurrecting Death: Anatomical Art in the Cabinet of Dr. Frederik Ruysch," *Art Bulletin* 1996;78:663–679.

18. www.famousquotes.me.uk/epitaphs/30.htm.

19. Anonymous, "A View of London and Westminster, 1728," cited in Ruth Richardson, *Death, Dissection, and the Destitute*, 2nd ed. (University of Chicago Press, 2000), 55.

20. Richardson, *Death, Dissection, and the Destitute*, 72.

21. Sappol M, *A Traffic of Dead Bodies: Anatomy and Embodied Social Identity in Nineteenth-Century America* (Princeton University Press, 2004), http://press.princeton.edu/titles/7178.html. Accessed Feb 26, 2011.

22. Bates AW, "'Indecent and Demoralising Representations': Public Anatomy Museums in Mid-Victorian England," *Med Hist* 2008;52(1):1–22.

23. *Times* (London), 19 Dec. 1873, p. 11.

24. Quoted in Jeffries S, "The Naked and the Dead," *Guardian.uk.com*, 2002, www.guardian.uk.com/education/2002/Mar/19/arts.highereducation. Accessed Feb 27, 2011.

Chapter One: Being Non-biodegradable

1. Burns L, "Gunther von Hagens' BODY WORLDS: Selling Beautiful Education," *Am J Bioethics* 2007; 7(4):12–23, and its commentary essays.

2. Starr J, "The Plastinates' Narrative," in *The Anatomy of Body Worlds: Critical Essays on the Plastinated Cadavers of Gunther von Hagens*, ed. Christine T. Jespersen, Alicita Rodriguez, and Joseph Starr (MacFarland, 2009), 8–15, at 11.

3. Stefan Hirschauer calls this the state of "'inorgorgs,' inorganic organisms" ("Animated Corpses: Communicating with Post Mortals in an Anatomical Exhibition," *Body and Society* 2006;12:25–52 at 26).

4. Ashton R, "Life after Death—My Future as a Skinless Wonder," *Independent on Sunday*, 6 July 2003.

5. Ibid.

6. Crace J, *Being Dead* (Picador, 2001), 100–101.

7. Lizama N, "Afterlife, but Not as We Know It: Melancholy, Postbiological Ontology, and Plastinated Bodies," in *The Anatomy of Body Worlds: Critical Essays on the Plastinated Cadavers of Gunther von Hagens*, ed. A. Rodriguez, T. C. Jespersen, and J. Starr (McFarland, 2008), 16–28, at 19.

8. Anonymous blogger, "*Body Worlds*, I Think I'm Dying," http://ithink imdying.wordpress.com/2008/01/21/body-worlds/ Web. Posted Jan 21, 2008. Accessed Nov 24, 2009.

9. Ashton, "Life after Death."

Chapter Two: Lifelike Humans

1. Press release, Zydel D, "Casino Royale's Cold War Symbols, Featuring Body Worlds," 28 October 2006.

2. von Hagens G, "Anatomy and Plastination," in *Body Worlds: The Anatomical Exhibition of Real Human Bodies* (Catalog on the Exhibition, 2006), 11.

3. Ibid., 31–32.

4. Wetz FJ, "Pushing the Envelope," in *Pushing the Limits: Encounters with Body Worlds Creator Gunther von Hagens on his 60th Birthday*, ed. Angelina Whalley (Arts and Sciences Books, 2005), 284.

5. Jones DG, Whitaker MI, *Speaking for the Dead: The Human Body in Biology and Medicine*, 2d ed. (Ashgate, 2009), 105.

6. Barilan YM, "Bodyworlds and the Ethics of Using Human Remains: A Preliminary Discussion," *Bioethics* 2006;20:233–247.

7. Annas GJ, *Worst Case Bioethics: Death, Destruction and Public Health* (Oxford University Press, 2010), 105.

8. Quoted in Danto AC, "Danger and Disturbation: The Art of Marina Abramovic," in *Marina Abramovic: The Artist Is Present,* ed. Klaus Biesenbach (Museum of Modern Art, 2010), 32 (emphasis in original).

9. Quoted in Nicol N, Wylie H, *Between the Dying and the Dead: Dr. Jack Kevorkian's Life and the Battle to Legalize Euthanasia* (University of Wisconsin Press, 2006), 17.

10. This argument was actually made to a California court that refused to grant a request that a terminal cancer patient be frozen while still alive. All agreed that cryopreservation, a long shot in the best of circumstances, would have a better chance of success if freezing was accomplished before death. (Italics are in the original quotation.)

11. Quoted in Regis E, *Great Mambo Chicken and the Transhuman Condition: Science Slightly over the Edge* (Addison-Wesley, 1990), 128.

12. Janofsky M, "Even for the Last .400 Hitter, Cryonics Is the Longest Shot," *New York Times,* 10 July 2002.

13. Shaughnessy D, "Open Question—Why?" *Boston Globe,* 8 July 2002.

Chapter Three: More Wondrous and More Worthy to Behold

Epigraph. As quoted in Sachiko Kusukawa, *The Transformation of Natural Philosophy* (Cambridge University Press, 1995), 105.

Chapter Four: Resisting the Allure of the Lifelike Dead

1. Lynch TT, "Tract," www.pbs.org/wgbh/pages/frontline/undertaking/undertakers/tract.html. Accessed Feb 28, 2011.

Chapter Five: Detachment Has Consequences

1. Treadway K, "The Code," *N Engl J Med* 2007;357(13):1273–1275.

2. Hafferty FW, *Into the Valley: Death and the Socialization of Medical Students* (Yale University Press, 1991).

3. Osler W, "Aequanimitas, Valedictory Address," University of Pennsylvania, May 1, 1889.

4. Fox RC, "Training for Uncertainty," in *Essays in Medical Sociology: Journeys into the Field*, ed. R. C. Fox (Transaction Publishers, 1988), 19–50.

5. Lewis CS, *The Abolition of Man* (HarperSanFrancisco, 2001), 72.

Chapter Six: The History and Potential of Public Anatomy

1. The anatomy theater was reconstructed after damage in World War II and can be viewed through a virtual tour at www.commune.bologna.it/girabologna/.

Chapter Seven: What Would Dr. William Hunter Think about
Bodies Revealed?

1. Sawday J, *The Body Emblazoned: Dissection and the Human Body in Renaissance Culture* (Routledge, 1995), 63.

2. Simmons SF, *William Hunter, 1718–1783: A Memoir*, ed. C. H. Brock (Glasgow University Press, 1893); Wilson A, *The Making of Man-Midwifery: Childbirth in England, 1660–1770* (Harvard University Press, 1995).

3. Gelfand T, "The 'Paris Manner' of Dissection: Student Anatomical Dissection in Early Eighteenth-Century Paris," *Bull History Med* 1988; 46(2):99–130, esp. 103.

4. Lawrence S, *Charitable Knowledge: Hospital Pupils and Practitioners in Eighteenth-Century London* (Cambridge University Press, 1996).

5. Simmons, *William Hunter*.

6. Richardson R, *Death, Dissection and the Destitute* (University of Chicago Press, 2001).

7. Ibid.

8. Ibid.

9. Simmons SF, Two Introductory Lectures, delivered by William Hunter, to his last course of Anatomical Lectures, at his Theatre in Wind-mill Street . . . Printed by order of the Trustees, for J. Johnson, London, 1784, pp. 113–114.

10. Whalley A, *Pushing the Limits: Encounters with Body Worlds Creator Gunther von Hagens* (Arts and Sciences Books, 2005).

11. Dobson J, *John Hunter* (E & S Livingstone, 1969).

12. Whalley, *Pushing the Limits*.

13. Simmons, Two Introductory Lectures, 7.

14. Whalley, *Pushing the Limits*.

15. Simmons, Two Introductory Lectures; Cazort M, *The Ingenious Machine of Nature: Four Centuries of Art and Anatomy* (National Gallery of Canada, 1997).

16. Jordanova L, "Gender, Generation and Science: William Hunter's Obstetrical Atlas," in *William Hunter and the Eighteenth Century Medical World*, ed. William Bynum and Roy Porter (Cambridge University Press, 1985), 387–412.

17. Simmons, Two Introductory Lectures, 55.

18. Dobson, *John Hunter*, 177.

Chapter Eight: Vive la différence

1. Tanassi L, "Plasti-Nation," lecture on 6 July 2006 at Weisman Museum, University of Minnesota, as part of the series: "The Body on Display: Controversies and Conversations," sponsored by the University of Minnesota Academic Health Center.

2. Stern M, "Shiny Happy People: 'Body Worlds' and the Commodification of Health," *Radical Philosophy* 2003;118 (March/April), www.radical philosophy.com. Accessed Feb 27, 2011.

3. Wolpert K, *Die Nackten und die Toten: Folgen einer politischen Geschichte des Körpers in der Plastik des deutschen Faschismus* (Anabas, 1982), 230.

4. O'Rorke I, "Skinless Wonders," *Observer*, Observer Review Pages, 20 May 2001, p. 5; Elliott C, "Human Taxidermy," lecture on 19 July 2006 at Weisman Museum, University of Minnesota, as part of the series "Stiff Morality: The Ethics of Using Bodies," sponsored by the University of Minnesota Academic Health Center.

Chapter Nine: Normative Objections to Posing Plastinated Bodies

1. "Jack" is a pseudonym used to protect privacy.

2. Photos of the *Bodies Revealed* exhibit are found at http://abcnews .go.com/Blotter/popup?id=4291499&contentIndex=1&page=1.

3. All biblical quotations are from the King James Version of the Holy Bible.

4. Hiscox ET, *The Star Book for Ministers* (Judson Press, 1968), 184.

5. McCabe J, *Service Book for Ministers* (McGraw-Hill, 1961), 125.

6. Lyrics by William Howe, 1864, public domain.

7. McCabe, *Service Book*, 153, 155.

8. Among contemporary adherents to the doctrine of soul-sleep are Jehovah's Witnesses and many Seventh Day Adventists.

9. Conklin BA, *Consuming Grief: Compassionate Cannibalism in an Amazonian Society* (University of Texas Press, 2001).

10. Spindler K, Wilfing H, et al., eds., *Human Mummies* (Springer-Verlag), 1996.

11. Gomez A, "Cemeteries Feel Recession's Chill," *USA Today*, 19 October 2009, p. 1A.

12. www.cremation.org/stats.shtml. Accessed Oct 23, 2009.

Chapter Eleven: The Creeping Illusionizing of Identity from Neurobiology to Newgenics

Epigraph. Wilner E, "Candied," in *Reversing the Spell: New and Selected Poems* (Copper Canyon Press, 1998), 227.

1. See, for example, the Symposium on Law, Ethics and the Historical Display of Human Remains, organized by Robert Juette and Christopher Lawrence (London: Wellcome Institute, April 13–14, 2005), www.ucl.ac.uk/histmed

2. The theological implications of destroying the human form to obtain knowledge are summed up in Goethe's distinction (in Wilhelm Meister) between the prosektor (the mutilator) and the proplastiker (the rejoiner). See Moore CM, Brown CM, "Gunther von Hagens and Body Worlds, Part I: The Anatomist as Prosektor and Proplastiker," *Anatomical Record (Part B: New Anatomy)* 2004;276B:8–14.

3. See my exhibition catalogue, *Depth Studies: Illustrated Anatomies from Vesalius to Vicg d'Azyr*, on loan from the Department of Special Collections, University of Chicago Library (Smart Museum of Art, 1992).

4. Haithman D, "Exhibition on the Human Body Gets under People's Skin," *Los Angeles Times*, 26 June 2004.

5. Hurtley S, Szuromi P, eds., "This Week in Science," *Science* 2005; 308(April 1):13.

6. Grene M, DePew D, *The Philosophy of Biology: An Episodic History* (Cambridge University Press, 2004), 291–303.

7. Morton O, "Life, Reinvented," *Wired,* January 2005, p. 173.

8. Ferber D, "Microbes Made to Order," *Science* 2005;303(January 9):158.

9. Shreeve J, "Craig Venter's Epic Voyage to Redefine the Origin of the Species," *Wired*, August 2004, pp. 149–150.

10. Thompson C, "How to Farm Stem Cells without Losing Your Soul," *Wired*, June 2005, pp. 118–120.

11. Black E, *War against the Weak: Eugenics and America's Campaign to Create a Master Race* (4 Walls 8 Windows, 2003), B.

12. Andrews LB, "Genes and Patent Policy: Rethinking Intellectual Property Rights," *Nature Reviews/Genetics* 2002;3(October):803–808.

13. Carlson R, "Splice It Yourself: Who Needs a Geneticist? Build Your Own DNA Lab," *Wired*, May 2005, pp. 89–90.

14. Lenoir T, "Shaping Biomedicine as an Information Science," in *Proceedings of the 1998 Conference on the History and Heritage of Science Information Systems*, ed. Mary Baldwin, Trudi Bellardo Hahn, and Robert U. Williams (Information Today, 1999), 43.

15. Pink DH, "Why the World Is Flat" (interview with Thomas Friedman), *Wired*, May 2005, pp. 151–153.

16. Orr HA, "Vive la Differance!" *New York Review*, 12 May 2005, p. 18.

17. Paradise J, Andrews L, Holbrook T, "Patents on Human Genes: An Analysis of Scope and Claims," *Science* 2005;307(March 11):1566–1567.

18. Germano W, "Passive Is Spoken Here," *Chronicle of Higher Education*, 22 April 2005, p. B20.

19. Zelder M, "Optimal Regulation of Genetic Testing: An Economic Analysis," presented at the Ethics, Genetics, and Pharmacogenetics Seminar, Maclean Center for Clinical Medical Ethics, University of Chicago, May 11, 2005.

20. Gooday GJN, *The Morals of Measurement: Accuracy, Irony, and Trust in Late Victorian Electrical Practice* (Cambridge University Press, 2004).

21. Lenoir, "Shaping Biomedicine as an Information Science," 27–28. Also see Doyle R, *On beyond Living: Rhetorical Transformations of the Life Sciences* (Stanford University Press, 1997).

22. Thackara J, *In the Bubble: Designing for a Complex World* (MIT Press, 2005).

23. See Taylor C, *Sources of the Self: The Making of the Modern Identity*

(Harvard University Press, 1989), and Kruks S, *Retrieving Experience: Subjectivity and Recognition in Feminist Politics* (Cornell University Press, 2000). Also see Ogbechie SO, "Ordering the Universe: Documenta 11 and the Apotheosis of the Occidental Gaze," *Art Journal*, Spring 2005, pp. 81–94.

24. Walter T, "Body Worlds: Clinical Detachment and Anatomical Awe," *Sociology Health Illness* 2004;26(4):28.

25. Rodman FR, *Winnicott: Life and Work* (Perseus Publishing, 2003), 185.

26. See my *Body Criticism: Imaging the Unseen in Enlightenment Art and Medicine* (MIT Press, 1991).

27. Thacker E, *Biomedia* (University of Minnesota Press, 2004), 5.

Chapter Twelve: Craft and Narrative in Body Worlds

1. LeWitt S, "Paragraphs on Conceptual Art," *Artforum*, June 1967.

2. www.feynmanlectures.info/stories/gustavo_duarte_stroy.html.

3. http://menmedia.co.uk/manchestereveningnews/news/s/1035155_bishop_blasts_body_snatch_show.

4. www.guardian.co.uk/world/2009/mar/10/venezuela-chavez-bodies-exhibition.

5. Bierce AG, *The Devil's Dictionary*, 1911 (definition of "Life"), www.gutenberg.org/files/972/972-h/972-h.htm.

6. Artaud A, *The Theater and Its Double*, trans. Mary Caroline Richards (Grove Press, 1958), 13 (original text 1938).

SUGGESTED FURTHER READING

Bertman S. *One Breath Apart: Facing Dissection*. Baywood Publishing, 2009.

Holman S. *The Dress Lodger*. Grove Press, 2010.

MacDonald H. *Human Remains: Dissection and Its Histories*. Yale University Press, 2006.

Montross C. *Body of Work: Meditations on Mortality from the Human Anatomy Lab*. Penguin, 2008.

Moore W. *The Knife Man: Blood, Body Snatching, and the Birth of Modern Surgery*. Broadway, 2006.

Richardson R. *Death, Dissection, and the Destitute*. University of Chicago Press, 2001.

———. *The Making of Mr. Gray's Anatomy*. Oxford University Press, 2008.

Rifkin BA. *Human Anatomy: From the Renaissance to the Digital Age*. Harry Abrams, 2006.

Rodriquez A and Starr J. *The Anatomy of Body Worlds: Critical Essays on the Plastinated Cadavers of Gunther von Hagens*. MacFarland, 2008.

Sappol M. *A Traffic of Dead Bodies: Anatomy and Embodied Social Identity in Nineteenth-Century America*. Princeton University Press, 2004.

Stafford B. *Body Criticism: Imaging the Unseen in Enlightenment Art and Medicine*. MIT Press, 1993.

Warner JH and Edmonson JM. *Dissection: Photographs of a Rite of Passage in American Medicine, 1880–1930*. Blast Books, 2009.

Worden G. *The Mutter Museum of the College of Physicians of Philadelphia*. Blast Books, 2002.

College of Physicians of Philadelphia, Mütter Museum in, 68, 122
consent, 2, 5, 30, 51, 79, 82–83, 86, 115
controversy, 3, 85–86, 91, 125, 127
copycats, 79, 81, 89
corpse: as art, 29, 85–86; authenticity of, 28; children's, 75; cryopreservation of, 33; dissection of, 97–98, 106, 120–121; factory and, 20–21; meditation on, 53; origins of, 5, 8, 10, 74; plastinated, 4, 13, 25–27, 39, 41, 65, 75, 90
Crace, Jim, 22–23
criminals, 50, 52, 56, 63–66, 74, 76
cryonics, 33
Cullen, William, 73
curiosity, 50, 60, 63
Curlin, Farr, 7, 15, 55–62

danse macabre, 119–120
Darwin, Charles, 107–108
death: acknowledgment of, 53, 55, 59; brain, 41, 44, 47; causes of, 18; funerary practices in, 19, 93–94; learning from, 34, 38, 40; reality of, 30–31; as sleep, 94, 96
decomposition, 2, 19, 23, 98
Denouës, Guillaume, 65
dignity, 2, 4, 30, 45, 85, 104
disease: anatomy of, 6; learning from, 33–34, 58, 122; venereal, 67, 69–70
display: animated, 47, 52, 65, 83, 93, 118; artistic, 9, 12–13, 75, 86; brain-dead bodies and, 42, 45; cryopreservation and, 33–34; ethics of, 30, 92; funerary practices and, 19; modes of, 60; plastination and, 1, 2, 4, 27, 39–40, 51, 56; posthumous, 4, 18, 23; as social issue, 73, 86, 92, 104
dissection: cadaver, 56–58, 61; embalming and, 2; medical education and, 71; opposition to, 7; public, 2, 8–9, 37, 50, 63–65

DNA, 20, 109, 111
donor: body, 18, 70, 84, 87–88, 95, 103–104, 112, 115; consenting, 79, 81, 83, 86, 91
Doyle, Richard, 111
Dr. Death, 80
Duchamp, Marcel, 116–117
Dürer, Albrecht, 39
Dying Gaul, The, 77

education: medical, 6, 15, 38, 71, 75, 92; moral, 90–91; public, 15–16, 37–38, 66, 69, 71, 82; science, 17, 30, 85; value of, 5–6, 12, 122–123
Egypt, 27, 39
Elliott, Carl, 87
embalming, 2, 19, 27, 74, 77–78
enlightenment, 3, 121
entertainment, 4, 31, 38, 40–42
ethics: abuse of, 82; Christian, 90, 99; concerns about, 3, 25, 41, 91, 103, 115; controversies in, 90, 93, 96–98, 124–128
Europe: controversy in, 2–3; dissection in, 7, 37, 63, 65; exhibitions in, 39, 81, 83, 85–86, 89, 111
euthanasia, 9, 125
evolution, 70, 109
exhibitions: anatomical, 27; art, 28, 84–85; controversy about, 2–4, 6, 12, 14, 32; educational, 12, 66, 69, 90; morality of, 80; museum, 1, 55, 81; of plastinates, 27, 37–44, 62, 125; Premier Exhibitions, 5–6, 79; religion and, 103

Faculty of Medicine, 64
Feinstein, Rabbi Morley, 4
Fellow of the Royal Society, 74
Feynman, Richard, 119
Florence, 63, 65, 72
Fox, Rene, 59
Fox Keller, Evelyn, 111
France, 7, 28, 39, 65
Frankfurter Allgemeine Zeitung, 80

freezing, 2, 31, 33–34, 106
funerary, 19, 93

genetics, 20, 43, 107, 109–111, 113
German Anatomical Society, 6
Germany, 4, 15, 26, 80, 88
ghouls, 14, 74
Gibbon, Edward, 74, 78
Giorgione: *Venus*, 86
Girl with the Dragon Tattoo, The, 26
Gonville Hall, College of Physicians at, 64
Goya, Francisco, 116, 118
Grass, Günter, 81
grave robbers, 10–11
Great Lakes Science Center, 81
Great Windmill Street, Soho, 74
gross anatomy, 61, 114

Hafferty, Fred, 57
Hanh, Thich Nhat, 53
Hansen, Julie, 9
Haraway, Donna, 111
Harvard Ad Hoc Committee on Brain Death, 40
Heaviside, John, 67
Holocaust, 81
Horse and Rider, 28, 49, 65
Hunter, John, 68, 76
Hunter, William, 68, 73–77
Hunterian Museum at the Royal College of Surgeons, 68
Hurlbut, William, 108
hybrid, 14, 20, 32, 46

Ingres: *Grand Odalisque*, 86
international trafficking, 91
Italy, 7, 65

Jardin du Roi (Paris), 64, 73
Jesus Christ, 94, 96–97
Jones, D. G., 6
Judeo-Christian ethics, 84

Kahn Anatomical Museum, 12, 67
Kant, Immanuel, 104, 121

Kevorkian, Jack, 32
King's Garden, 73
Kurtz, Steve, 31

Lantos, John, 1–16, 124–128
Lascaux, France, 28
La Specola, Florence, 65, 70
Leiden, 9, 13
Lenin, Vladimir, 45
Lenoir, Timothy, 109, 111
Lewis, C. S., 62
LeWitt, Sol, 116–117
Lizama, Natalia, 23
London, 11–12, 64, 68, 84, 116
London Company of Barber Surgeons, 64
Lynch, Thomas, 51

Ma, Fiona, 80
Manchester Museum of Science and Industry, 119
man-midwifery, 73–74
Mapplethorpe, Robert, 85
McCulloch, Right Reverend Nigel, 119
meat market, 81
Melanchthon, Philipp, 37
memorials, 61
Mengele, Joseph, von Hagens compared to, 81
Miller, Jonathan, 43
models: diseases in, 12; plastinate, 14, 20, 53, 86; wax, 65–67, 107
Mondino dei Liuzzi, 63
Montross, Christine, 15, 48–54
mummies, 45–46, 135
Munich, 2, 81
Muscle Man, 27
Musée Fragonard d'Alfort, 39, 65
Museum of Modern Art (MoMA), 31
Museum of Science and Industry (Chicago), 49, 82, 122
Mütter Museum (Philadelphia), 68, 122
Mystical Plastinate, 112